J. K. Rassweiler, Henry H. Rassweiler

The Teachers' Manual and Pupils' Text Book

Anatomy, Physiology and Hygiene

J. K. Rassweiler, Henry H. Rassweiler

The Teachers' Manual and Pupils' Text Book
Anatomy, Physiology and Hygiene

ISBN/EAN: 9783744670043

Printed in Europe, USA, Canada, Australia, Japan

Cover: Foto ©berggeist007 / pixelio.de

More available books at **www.hansebooks.com**

THE TEACHERS' MANUAL

AND

PUPILS' TEXT-BOOK

ON

Anatomy, Physiology and Hygiene,

INCLUDING THE EFFECTS OF

ALCOHOL AND NARCOTICS UPON THE HUMAN SYSTEM,

DESIGNED TO ACCOMPANY

THE TEACHERS' ANATOMICAL AID

BY

PROF. J. K. RASSWEILER, A. M.

AND

PROF. H. H. RASSWEILER, A. M.

REVISED AND ENLARGED EDITION.

PUBLISHED BY

CENTRAL SCHOOL SUPPLY HOUSE,

CHICAGO, ILLINOIS.

PREFACE.

This book is intended to go into the schools of the country as a companion to the "TEACHERS' ANATOMICAL AID," which is supposed to be before the class during every exercise or recitation in Physiology, and to which the illustrative references, found throughout these pages, relate.

To the teacher who gives *oral lessons* in Physiology, by the use of the Aid, this volume offers assistance in presenting the truths of the science in proper order, plain language, and with many illustrations gathered within the range of the pupils' observation and experience. Thus, even inexperienced teachers are furnished with methods and material to conduct a well-arranged course of daily drills on a subject of surpassing importance and interest.

To the teacher who conducts a *recitation*, with the use of the Aid, this work offers guidance in pointing out, precisely, by its frequent references, those parts or features on the plates or manikin sections, which illustrate any topic in hand, as found in the current text-books on Physiology.

While this book is thus intended to be helpful to

teachers of all grades of experience, it is, at the same time, adapted for use as a text-book for elementary classes. The practical results which may be obtained from such a use of it, in connection with the Anatomical Aid, will be found to be more satisfactory than those which can be attained by any other method of instruction in the elements of Physiology.

When used as a text-book with the Aid, the latter should be made conveniently accessible to the pupils. This can easily be done in any schoolroom. The objection that the pupils will injure the charts by handling them, is a mischievous notion. They are entitled to such privileges. If well and kindly advised, they will handle them properly and will take pride in carefully preserving them from injury.

TABLE OF CONTENTS.

4   CONTENTS.

THE SKELETON.

BONES.

Like the Framework of a House. The framework of the body is composed of *bones* and *ligaments*. It is called the skeleton. What beams, joists and rafters are to a house, bones are to the body. As each timber in the framework of a building is fitted for its own particular place and purpose, so each one of the bones of the body has its own place and is in every way precisely adapted, in shape and strength, for a special use.

Number of Bones. There are two hundred and eight bones in the skeleton. This does not include the teeth, for they are really not a part of the skeleton. Thirty-four of the bones of the body are single—only one of the same kind. Besides these, there are eighty-seven pairs, the two bones of each pair being alike—one on each side of the body.

Shape. The skeleton plate shows that the bones are very different in shape. Some are long, like this (leg bone), for example. Others

(5)

are nearly round, like these bones of the wrist. Some are quite flat, like this large, spreading bone at the shoulder (16), or these broad bones in the lower part of the main body (3).

Structure. The bones are very hard and strong. "Hard as a bone" is a familiar comparison. We shall not be surprised at their hardness and strength when we shall have learned what important uses they serve in the body. There are two kinds of material in the structure of a bone. One part is called *animal matter* and the other is called *mineral matter*. If the bone were composed of animal matter alone, it would bear no pressure and keep no permanent shape. If it were made entirely of mineral matter it would be too brittle, and consequently would break very easily. So these two kinds of material are united together in such a way as to secure strength without too great brittleness. In childhood the bones are not easily broken. This is because in early life they contain about twice as much animal matter as mineral matter. What a wise protection against the "bumps" and "tumbles" of the little ones. In middle life the two kinds of material are more nearly equally divided. In old age, however, the bones are very brittle, because, then, there is about twice as much mineral matter as animal matter in their structure.

It is an easy and interesting experiment to separate these two kinds of bone material. Throw a flat bone, or piece of bone, into the fire. After a while you will find a part of it, like a cinder, among the ashes. This is the mineral part. The fire has burned out the animal matter. Take the "drum-stick" bone of a chicken and place it in a bottle containing a mixture made by filling the bottle half full of water and adding about half as much muriatic acid—a common drug which you can get for a trifle at the nearest drug store. This will take out the mineral matter from the bone and leave the animal matter. The mineral matter which you took from the fire was brittle or crumbling. The animal matter, when taken from the acid, is gluey and can be wound, like a cord, about the finger. The broad or flat bones, like those of the head, are not entirely solid. Between the two outside layers of such a bone there is a layer of spongy-like material. These three layers of structure in a flat bone can be clearly seen by looking at the edge of such a bone which has been sawed through.

The long bones are generally hollow and contain a substance called *marrow*. At the ends they are usually thicker and more spongy. This serves to break the force or shock of heavy stepping or jumping with the lower limbs, or a hard stroke with the arm. The ends of the long bones are also covered with a smooth, white substance called *cartilage*. This aids in giving the bone an easy motion at the joint where it is united to another bone.

How United. The bones are united to each other in different ways. Those which are quite movable are connected by joints. Some of these are called *hinge-joints* because they work like the hinge of a door. These (arm) bones which meet at the elbow are hinge-jointed. Raise and lower your forearm and notice particularly how the joint acts. The joints in the fingers and the knee are also hinge-joints. Another kind is called the *ball and socket joint*, where the round end of one bone moves in a hollow place of another. Here (*a*) at the hip is a good example of a ball and socket joint, where the round head of this large upper bone of the leg moves in a deep hollow of this lower bone of the main body. The bones of the head meet each other with jagged edges forming a seam-like junction called a *suture*. One of these is clearly shown on this skeleton (12). Between the bones of the back are placed cushions of *cartilage*. This is a substance softer than bone and quite elastic, like rubber. This cushion arrangement between the bones of the back, is nicely shown on this plate. (Refer to cartilages between lumbar vertebrae.)

Bound To-gether by Ligaments. The bones are bound to one another by *ligaments*. These are very strong and hold the bones firmly in position. Some of these stout bands or ligaments are shown on this plate. Here (XVI) are the ligaments which bind together the bones of the hip. These (XXV, XXVI) are the ligaments of the elbow joint.

Uses of the Bones. The bones of the body serve several important purposes. 1. They give the body its general shape. 2. They support the softer material of the body within and around them. 3. They protect delicate and important parts against injury from without, as, for example, the brain, lungs and heart. 4. They serve as levers, to be moved by the muscles in the various movements of the body, as we shall learn more clearly, somewhat later.

Divisions of the Skeleton. By looking at the figure of the skeleton we perceive that the bones are grouped into four natural divisions, namely: 1. The bones of the *head*. 2. Those of the main body, or *trunk*. 3. Those of the *upper extremities*, or arms. 4. Those of the *lower extremities*, or legs.

We have now learned about the number, shapes, material, union, uses and groups of the bones of the skeleton. We are now ready to study the more important bones of each group more closely.

TABLE OF THE SKELETON.

Bones whose proper names are not given elsewhere are printed in *italics*. Letters and figures in parenthesis refer to the bones as shown in Anatomical Aid.

BONES OF THE HEAD.

1 Frontal (forehead)............................(1)
2 Parietal (upper side of head)................(2)
2 Temporal (lower side of head)................(3)
1 Occipital (back of head)......................(4)
1 *Sphenoid* (base of skull).
1 *Ethmoid* (base of skull).
2 Nasal (bridge of nose)........................(6)
2 *Malar* (cheek bones)........................(4)
2 *Lachrymal* (part of eye socket).
2 *Palate* (roof of mouth).
2 *Turbinated* (in cavity of nose).
2 Upper Maxillary (upper jaw)..................(7)
1 Lower Maxillary (lower jaw)..................(8)
1 *Vomer* (partition between nostrils).

BONES OF UPPER EXTREMITIES.

2 *Clavicle* (collar-bone)......................(8)
2 *Scapula* (shoulder-blade)..................(16)
2 Humerus (upper arm)..........................(1)
2 Radius (forearm)..............................(2)
2 Ulna (forearm)................................(3)
16 *Carpal* (wrist bones)......................(4)
10 *Metacarpal* (middle hand)..................(5)
28 *Phalanges* (finger bones)...............(6–10)

BONES OF THE TRUNK.

7 Vertebræ of the neck......................(1–3)
12 Vertebræ of the back.....................(4–5)
5 Vertebræ of the loins.....................(6–7)
14 True Ribs...............................(12–13)
10 False Ribs.............................(14–15)
1 Sternum (breast-bone)......................(10)
2 *Innominate*, or Hip Bones..................(3)
1 Sacrum (between hip bones)..................(1)
1 *Coccyx* (cuckoo bone).

BONES OF LOWER EXTREMITIES.

2 Femur (thigh bone)...........................(1)
2 *Patella* (knee-pan)........................(2)
2 Tibia (leg bone)............................(a)
2 Fibula (leg bone)...........................(b)
14 *Tarsal* (heel and instep).
10 *Metatarsal* (middle foot).
28 *Phalanges* (toe bones).

The above, with the *Hyoid* bone, and four bones in each ear, makes 208 bones.

BONES OF THE HEAD.

The Skull. There are twenty-two bones in the head. Eight of these are shaped and united in such a way as to form a sort of round box which is called the *skull,* or *cranium.* This is one of the most important parts of the skeleton, since it contains the brain, the most delicate organ of the body. The word *organ,* in physiology, means any single part of the body which serves a special purpose. Thus, the brain, heart, lungs and veins are organs. The skull, or brain-box, is placed, like a dome, at the top of the structure of the body. It is wonderfully fitted for the protection of its tender contents. It is shaped for strength as well as for beauty.

The front of the skull is formed by the *frontal* bone (1), or bone of the forehead. The two *parietal* bones (2) form the upper sides, and the two *temporal* bones (3) form the lower sides of the skull. At the back of the head (4, back view of skeleton) is the *occipital* bone. Two more of the skull bones form its lower part or base. These we will not name here; but you will find them named in the full table of the bones which has been given.

The Face. The remaining fourteen bones of the head give shape to the face. The two *nasal* bones (6) form the bridge of the nose, and the two *malar* bones (4) the prominence of the cheeks. The upper jaw is formed of the two *upper maxillary* bones (7). The lower jaw bone (8) is called the *lower maxillary.* The teeth are set in sockets of these maxillary bones.

BONES OF THE TRUNK.

The main body is called the *trunk.* The upper part of the trunk is fitted to contain the lungs and the heart. Its lower part contains the stomach, liver and bowels.

The Spinal Column. The great pillar of the body is the *spinal column.* It bears aloft the head—the crowning part of the whole structure. It supports the great vital organs of the main body. It is most wonderfully constructed with reference to comfort and safety of life. Instead of being composed of but one or a few bones, it is built up of twenty-six pieces, which, while laid up one above the other, are separated from each other by very elastic cushions of cartilage. This does not only make the back-bone capable of bending forward, backward and sideways, but it makes the whole pillar springy, so that the delicate brain which rides at its summit is not affected by jarring from the heavy movements of the body.

Vertebræ. Twenty-four of the bones of the spinal column, or back, are called *vertebræ* These are firmly bound together by ligaments and interlocked with each other by their own projecting parts. An opening runs through each vertebra. These openings form the *spinal canal* through which

the *spinal cord*, of which we shall learn hereafter,
passes. The vertebræ are divided into three sets. The
seven upper ones are in the neck. The next twelve
are in the back proper; to these the twelve pairs of
ribs are attached. The five lower vertebræ are in
the region of the loins. They are very stout, as the
figure shows, just as we would expect them to be,
since they support a large part of the weight of the
body.

The Chest. The upper part of the trunk, which contains the heart and lungs, is called the
chest. The skeleton of the chest is formed behind,
as you see, by the middle division of the spinal column; on the sides by the *ribs* (12, 13), and in front
by the *sternum* (10) or breast-bone. Here, again, we
find a wise provision for the protection of life. The
breast-bone is not near so hard as most of the other
bones. It is consequently more flexible. The ribs
are not directly united to the sternum, but are joined
to it by cartilages. By these means, a heavy blow
on the breast, which otherwise would seriously injure
the organs within, is made comparatively harmless.

The plate of the skeleton shows that the ribs are
not all joined to the breast-bone in front. Seven
pairs (12-13) are so joined. These are called *true
ribs.* The remaining five pairs (14-15) are called
false ribs.

The Pelvis. The bones of the lower part of the trunk
are shaped and joined so as to form a
large bowl-shaped cavity. This is called the *pelvis.*

Notice how broad and peculiarly formed these two (3) bones are. The *sacrum* (1) bone is wedged between these two bones at the back.

BONES OF THE UPPER EXTREMITIES.

The Shoulder. In examining the shoulder, we first notice these two collar-bones (8). Their use is to brace the shoulders properly apart; so one end rests against the breast-bone and the other against the shoulder. Next comes the shoulder-blade (16). These are so broad to allow the attachment of some very strong muscles of the upper part of the body.

The Arm. The upper arm has one large bone called the *humerus* (1). The *radius* (2) and the *ulna* (3) are the bones of the lower arm. There are eight roundish little bones in each wrist (4), five longer ones in the middle of each hand (5), three short bones in each finger (6, 9, 10) and two in each thumb (7, 8).

BONES OF THE LOWER EXTREMITIES.

The Thigh. Here we meet the largest bone of the skeleton (1). It is called the *femur*. Observe the round shape of its upper end (*a*). This is called the head of the femur. Moving in a hollow place of this large bone (3), it forms the ball and socket joint of the hip. Do not fail to notice how securely the lower limbs are bound to the main body by these numerous strong ligaments.

Lower Leg and Foot. The lower leg has two bones—the *tibia* (*a*), and the *fibula* (*b*). The knee-joint between the thigh bone and the bones of the lower leg, is protected by a flat bone called the knee-pan (2). There are seven bones in each heel, five in the middle part of each foot, three in each of the larger toes, and two in each great toe.

Notice this peculiarity in the form of the foot. It is curved or arched from the heel to the front. Here is another arrangement for springiness, without which, walking would not only become clumsy but painful.

Health of the Bones. The bones of a grown person are so much hardened by the mineral matter which has increased in their structure, that they are not easily changed in shape. They are more easily broken than bent. Neither is the full-grown joint likely to change in shape and character during the active years of life. So the general carriage of the body in adult life, depends on the habits and circumstances which shaped it in youth. We have learned that in childhood the bones are quite flexible and disposed to bend, instead of breaking, under a strain. For this reason, children who begin to walk very early become more or less bow-legged. Pupils who are in the habit of leaning forward on the desk, in school, will certainly, more or less deform their bodies. If a boy, in walking, carries his body in a lazy, stoop-shouldered position, he will go bent and deformed through life. Avoid leaning the body forward in sitting. When lying down, do not bolster up the head with high pillows. While standing or walking, hold the head erect, throw the shoulders back, and take in full breaths of air. If these positions in lying, sitting, standing or walking are carefully kept in youth, all the curvings and efforts of the body and limbs which come from ordinary labor, will not injure them, and the full-grown figure will be straight, graceful and strong.

OUTLINE.

THE BONES OF THE SKELETON.

WHAT?
- Framework of the body.
- 208. Eighty-seven pairs. Thirty-four single.
- Shapes: Long, flat, round, irregular.
- Composed of animal and mineral matter.
- United by joints, sutures and cartilages.
- Bound together by ligaments.
- Arranged in four groups:
- Head, Trunk, Upper and Lower Extremities.

WHERE?
- Twenty-two in the head.
- Fifty-four in the trunk.
- Sixty-four in the upper extremities.
- Sixty in the lower extremities.
- Eight in the ears.

WHY?
- To give shape to the body.
- To support the softer parts of the body within and around them.
- To protect important organs.
- To serve as levers to be moved by the muscles.

SUGGESTIONS TO THE TEACHER.

Be sure that the acid and burning experiments on the *composition* of bones are performed either by yourself or by the pupils. Get a piece of flat bone sawed to show the *layers*. Get a leg joint at the butcher's; remove muscles and tendons, to show the *ligaments;* then sever the bones at the joint to show *cartilage.* Show a fresh piece of long bone containing *marrow.*

TEST QUESTIONS.

Of what is the skeleton composed?
To what parts of a house are the bones compared?
How many bones in the skeleton?
Do the teeth belong to the skeleton proper?
How many single bones in the body?
How many are in pairs?
What variety of shapes have the bones?
What two kinds of material in the bones?
Which material makes the bone flexible?
What is the effect of the mineral matter?
How do these materials vary at different ages?
What wise provision in this arrangement?
Why are the ends of the long bones more spongy?
Why are they covered with cartilage?
In what ways are the bones united?
Locate a hinge-joint of the body.
Where is a ball and socket joint found?
What bones are united by sutures?
What bones are united by cartilages?
How are the bones bound to each other?
Name four uses of the bones.
Into how many groups are the bones divided?
What is the skull?
Point out on the Aid, the frontal bone,—the parietal—temporal—occipital.
How many bones form the face?
What two form the bridge of the nose?
Where are the malar bones?

The upper maxillary?   Lower maxillary?
What is meant by the trunk?
Where is the spinal column?
Why is it built so strong?
How is it made, elastic or springy?
What benefit in this arrangement?
How many vertebræ in the back-bone?
How many of these are in the neck?
How many have ribs attached to them?
How many are in the loins?
What bones make the frame of the chest?
Are the ribs united directly to the sternum?
Is the sternum as hard as other bones?
What benefit in these arrangements?
How is the lower part of the skeleton of the trunk shaped?
What is it called?
What three bones come together at the shoulder?
What two in the forearm?
How many in the wrist?
How many in the middle of the hand?
In each finger?
In each thumb?
Which is the longest bone in the skeleton?
What two bones in the lower leg?
How many heel bones in each foot?
Why is the foot arched instead of flat?

THE MUSCULAR SYSTEM.

MUSCLES.

We have studied the framework or skeleton of the body. We have seen from the figure of it in the Anatomical Aid, how it resembles the framework of a house before it is weather-boarded and shingled. The plate of the body which is now before us presents altogether a different view from that which we have been studying. We notice that it looks more like the full body, more like a house that is enclosed. The bones are here quite concealed by another division of organs—the muscular system. The word *system* in Physiology means the whole collection of parts or organs of the body, which perform similar work or which work together for some common purpose. We are already acquainted with the bony system. We will now study the muscular system.

The *muscles* form the lean flesh of the body. The meat which we eat for food is chiefly muscle. We are all familiar with the dark red color of beef when it is

The Muscles.

(19)

raw. You have also undoubtedly noticed that the muscle or lean meat of pork is of a paler red, and the meat on the breast-bone of a chicken is quite white; so muscle is not always red; but it is *generally* red, and the plate shows us that the muscles of the human body are of a quite red color.

Number and Structure of Muscles. There are 527 muscles in your body. Each one of these is made up of many strands or string-like fibers. These are laid side by side in the muscle, sometimes making quite a thick bundle. Each fiber of a muscle bundle is, however, separated from the rest by a very delicate substance. If you will take a piece of cooked meat, when it is cold, you can pull the muscles apart into strands, and these strands can be separated into many finer fibers or threads of muscle. While this is being done, you can observe the breaking and crackling of the very thin layer of matter which separates the fibers. The muscles differ from each other in shape. Some are spread out much like a fan. Others are quite circular in form, like this one (5) around the eye, or this (15) around the mouth. Some are quite long and of nearly even thickness. The largest muscle in the body is this (60), called the tailor muscle. It is nearly a yard long and does the work of crossing the legs.

Tendons. The ends of the muscles are attached to the bones by means of a hard white substance or cord, which is called a *tendon.* These tendons are very strong. Besides binding the muscles very firmly at their ends to the bones, they are very useful in giving a graceful shape to many parts of the body. For instance, if these (39 and 40) muscles of the forearm, which must have a connection with the fingers, were all continued as muscular bundles, through the wrist, hand and finger-joints, the hand would have a very clumsy figure. But these muscles reach out to the finger-joints by means of their tendons, and these tendons are neatly bound down, to run snugly along the bones, by means of ligaments, like this (45), so that the hand is really a very shapely organ. This (63) shows the tendon of this (62) muscle of the leg, and here (68) is the tendon of this (66) large muscle of the thigh.

Use of the Muscles. The muscles have been very appropriately called "our servants," furnished us with "the house in which we live." They are indeed very faithful servants. It is their work to move, in many ways, the different parts of the body; or, as in walking, to move the body as a whole. There is no movement of any part of the body which is not produced by the action of one or more muscles. Every step we take, the slightest motion of a finger, the movement of the lips in speaking, the chest in breathing, or the eye in winking — all these movements are produced by the muscles. The rapidity with which these muscles work is quite astonishing. To be convinced of this, we may observe the movements of the fingers of a skillful pianist or a rapid type-writer. To help you understand still better how very rapidly the muscles can act, you may remember that in saying the one word *muscle*, the mouth, tongue and voice organs must be put, in succession, into four different shapes or positions, all of which is done by the proper muscles. We must not get the idea that only the bones are moved by the muscles. Many other parts of the body are moved by their action. For instance, the lips in whistling, the eye-lids in winking, the skin in wrinkling the forehead, or the heart in its ceaseless beating. When a dog pricks up his ears, or a horse drives off the flies by shaking his skin, it is done by the action of the muscles.

Language of the Muscles. There is another use which the muscles serve, which is very interesting. It may be called the *language* of the muscles, and it is remarkable how often they speak for us. A frown on the face is purely the work of the muscles; yet everybody understands its meaning. The same is true of a smile. You see two men at a distance standing face to face and near together, with clenched fist and up-raised arm. You do not hear a word they say, but the action of their muscles, which you see, tells you how they feel. You pass near by a vicious horse, as he lays back his ears, or approach a dog whose hair on his neck is drawn up stiff and straight, you hear no sounds, but you understand the warning. It is the silent but expressive language of the muscles.

Two Kinds of Muscular Action. Some of the muscles of the body act only when they are directed to do so by the mind or the will. These are called *voluntary* muscles. Others act without being controlled by the will. These are called *involuntary* muscles. The muscles of the arm, for example, are voluntary muscles. The muscles which produce the action of the heart are involuntary. Some muscles may act either with or without the action of our will. For instance, the muscles which produce winking usually "wait for no thinking." But we may will to wink, and wink whenever we please. On the other hand, the will usually controls the action of the motion of the jaws. But sometimes, as in the case of a chill, these muscles produce chattering of the teeth rather contrary to the direction of the will.

How the Muscles Act. Motion is produced by the muscles, by the *contraction* of the fibers. A muscle shortens more or less according to the degree of motion which it is to produce. The shortening in length is caused by a swelling out of the muscles sideways. This swelling or bulging of a muscle can easily be perceived while it is contracted and pulling or holding the part which it moves. Grasp your arm between the elbow and shoulder firmly between your thumb and fingers. Now raise your forearm toward your shoulder; you feel the thickening of the muscle which raises your arm. This (34) is the muscle whose action you so plainly feel. It is called the *biceps* muscle of the arm. This

name means double-headed, and this muscle is so
called because it has two upper tendons or starting
places. Here (32) is the one, and here (33) is the
other. The return of the muscle to its usual shape
and length is called its *relaxation.* The relaxation
of this (34) biceps must take place to permit the
arm to straighten out; but, at the same time, some
other muscle or muscles must contract to move it
into the straight position. A muscle which bends a
part is called a *flexor.* One which serves to
straighten a part is called an *extensor.*

Antagonists, or Counter Muscles. Most of the muscles of the body are
paired off in their work. That is, the
motion of a part produced by a certain
muscle is reversed by the contraction of some other
muscle. Such muscles are called antagonists, or
counter muscles. Here again we refer to the chart
for illustration. To raise the forearm, as we have
seen, this (34) biceps must contract; but to straighten
it out again requires the action of this muscle (36),
the triceps. So the biceps and triceps are antagonists. These muscles (43 and 44) bend the fingers,
while these (51 and 52) straighten or extend them—
another illustration of counter muscles.

Some Prominent Muscles and Their Names. The names of the muscles are very
long and difficult to remember. It
would be unwise and unreasonable
to ask you now to learn many of them. But by
studying a few of the more prominent ones you will
learn something about their uses, and also how their

names are formed. This (1) muscle, which occupies a very prominent place, begins on the occipital or back bone of the head, and reaches forward to the skin of the forehead over the frontal bone. Its contraction raises the eyebrows and wrinkles the forehead. It is called the *occipito-frontalis.* It takes its name from the parts which it connects. This (15) curious muscle, when it contracts, puckers the lips. Physiologists call it *orbicularis oris.* Orbicularis means *circular,* and oris means *of the mouth.* So this muscle is named from its shape and position. Here (51) is a muscle which bears the name *extensor indicis,* which means the straightener of the index finger, this being precisely the work which the muscle performs. This muscle (22) takes its name from its position under the clavicle or collar-bone. So it is called the *sub-clavian* muscle. We see that some muscles are named after the parts which they connect; some from their shape and position; some from the work which they do, and others from their location. So the many long and difficult names of the muscles which you find on this plate (to which the figure seems to be pointing), and which are so meaningless to you now, are really very expressive and full of meaning, and may, some day, when you are more advanced in your studies, become very interesting to you.

Health of the Muscles. The comfort of the body, its grace of form and the prompt activity of all its parts depend very largely on the healthy and vigorous condition of all the muscles. To keep them all in that condition, each one must be used without being abused. A muscle which is not used loses its power of contraction, becomes weak and flabby, and finally altogether useless. On the other hand, if a muscle is overworked, it loses its power. If you were to tie up your arm in a sling, or bind it down to your side for a long time, you would lose the use of it entirely. If you should swing your arm for a long time, the muscles which produce its motion would cry out in painful protest against the abuse which they suffer; and were you to disregard their protest, they would "strike" and refuse, positively, to do the bidding of your will. The effect of the vigorous exercise of the muscles without overtaxing them, is to make them firm and strong; the stout arm of a blacksmith, and the strong limbs of a footman illustrate this. The difference between the robust figure and good health of a sturdy country boy and the slender body and feeble strength of his young friend in the city, lies mostly in the difference in amount of their general muscular exercise. But we must be careful not to mistake a bulky body, or thickness of the limbs, as a sign of stoutness and strength of muscle. It is true, indeed, that as the muscles grow stronger they grow thicker, and consequently increase the size of the limbs and trunk of the body. But the effect of the *fat* of the body is often mistaken for an "abundance of muscle."

OUTLINE.

THE MUSCLES.

WHAT?
- The lean flesh of the body.
- Color, red. Number, 527.
- Composed of many fibers.
- Shapes : long, fan-shaped, flat and circular.
- Bound to the bones by tendons.
- Voluntary and involuntary.
- Have power of contraction.
- Swell out when they shorten.
- Antagonists produce counter motion.
- Flexors bend, extensors straighten.
- Are kept healthy by exercise.

WHERE?
- Found distributed in all parts of the body.

WHY?
- To give motion to all parts of the body by the contraction and relaxation of their fibers.

SUGGESTIONS TO THE TEACHER.

In these lessons, whether you teach them by oral exercises or in recitation by the pupils, you can add much interest and practical instruction by bringing before your class illustrations of the real working of the parts or organs which are being studied. This can often be done very conveniently, and will contribute much to the pupils' knowledge of the *functions* or use of the organs (physiology), while the Anatomical Aid gives them a correct view of the *structure* (anatomy) of the parts. In studying the muscles, especially, such real examples of their work are very easily given. Name and point out on the plate a certain muscle. Make it serve your will as your pupils look on. Then let the class, in concert, join you in the performance. Wrinkle the forehead, close the eyes, pucker the mouth, swell the cheeks, raise the arm, etc. This will make the information which is imparted "stick," because it is stored in the mind among the *pleasures* of memory.

The "drum-stick" of a chicken—which some pupil may like to contribute—will, at this stage, furnish a very good object-lesson. Show how the muscles are grouped about the upper part and gradually taper down to the bone. Below the muscle, lying along the bone, is a tendon. Separate the muscle. If the "drum-stick" has become cold, after having been cooked, you may hear the crackling of the delicate little sheaths which encase the fibers. When you have removed all the muscles, you have left two representatives of the bony system—the larger bone, the tibia, and the slender bone by its side, the fibula, corresponding, in position, to the same bones in the human body.

TEST QUESTIONS.

What part of the body do the muscles form?
What is the usual color of the muscles?
How many muscles are in the body?
What can you tell of the structure of a muscle?
How do muscles differ in shape?
What is the shape of the muscle which closes the eye?
What and where is the longest muscle of the body?
What is the use of the tendons?
Can you explain how the tendons assist in giving a graceful
    shape to the body?
What is the use of the muscles?
What parts, besides the bones, are moved by the muscles?
Can you give an illustration of the language of the muscles?
What is meant by a voluntary muscle?
What is an involuntary muscle?
What is meant by the contraction of a muscle?
What by relaxation?
What is the difference between a flexor and extensor muscle?
What are antagonists or counter muscles?
How is a muscle affected by being unused?
What is the result of too severe exercise?
Is a bulky body always a strong body?
What is likely to make the body bulky?

THE NERVOUS SYSTEM.

So far as we have now studied the body we have its framework and the muscles which are to give motion to its various parts. We have learned *how* the muscles act, and now comes the question: What causes them to act as they do? We have learned of the obedience of the voluntary muscles to the will. But how does the mind or will direct them *when, how much*, and *how long* to act? For the purpose of enabling the mind to control the action of the muscles, a very interesting system of organs is provided in the body, namely, *the nervous system.* This plate gives us an excellent view of it.

The Brain. The *brain* is, in many respects, the most important organ of the body. It occupies the loftiest chamber of the body house. (*Raise the face section and refer to the brain on plate.*) Here the mind—the invisible tenant or occupant of the body—seems to form its purposes and send out its orders to its hundreds of servants stationed at as many points, between top and toe. Here, also, it receives its messages of intelligence from the body and from the outside world. These messages may bring it pleasure or pain; and they largely influence its decisions, its orders and its temper.

Structure. The brain is an exceedingly soft and delicate organ. If it were not enclosed in a triple sac and nicely fitted into its bony chamber it would fall apart from its own weight. It is composed of two kinds of substance, one of which is *gray* in color and the other *white.* The outer portion of the brain is composed of the gray matter. The white matter occupies the inside portion.

Protection to the Brain. The brain is surrounded by three coats or membranes. The one lying next to it is a delicate covering containing vessels which supply the brain with blood. This membrane takes its name from its purpose of careful protection; so it is called the *pia mater* — which means *a tender mother.* It lies very close to the surface, stretching over the little hills and dipping down into the little valleys, with which the outside of the brain is covered. Next to the *pia mater* lies a membrane so delicate that it was named after a spider's web—*arachnoid.* This membrane performs its work of protection by collecting from the blood a watery fluid to moisten the surface of the brain and prevent any possible friction. The outer coat is quite tough and substantial: so it is called the *dura mater,* or hard mother. It lies close to the inside surface of the skull bones. Now we can see how the brain is protected, for instance, against a blow on the head. The effect of such a blow would be diminished, first, by

the hair, then by the skin and muscles overlying the
skull, then by the bone, next by the hard coat, then
by the water coat, and finally by the soft coat—making no less than a half-dozen successive defenses
against harm to the castle of the mind.

Divisions of the Brain. The brain is divided into two parts, one
of which is much larger than the other.
These parts are shown here, in this
section which represents the head as divided from
top to bottom, close behind the ears. We will now
refer to the manikin of the head, where we will get
a very clear view of the size and position of these
brain parts. (*Fourth section of the head.*) This
(74) large upper brain is called the *cerebrum*. It
fills the whole front and upper part of the brain-box.
The small brain (75) is called the *cerebellum*.
Notice that it lies behind and below the large division of the brain. When this small brain is cut
through, its inner structure has this tree-like appearance (*shown on plate*), called the *arbor vitæ*.

Hemispheres. Both the cerebrum and cerebellum
are divided into two parts, called the
right and left *hemispheres*. The lower parts of the
two hemispheres are united by several small mys-
terious-looking organs, whose particular use has
been a puzzle even to many wise heads, but which
certainly have some special part to perform in the
wonderful control of the mind over the body. The
last section of the head (*turn to it*), which represents it as cut through from front to back, in the

middle, shows us the right hemisphere of both the larger and the smaller brain. The red vessels, in the figure, are *blood-vessels* which bring large quantities of the purest blood in the body to the brain, for a purpose of which we shall learn hereafter.

Work of the Cerebrum. From many observations and experiments which have been made by physiologists, it has been learned that the large brain is the *thinking organ* of the mind. It is here that impressions received from the outside world are translated into thought and feeling. Here the purposes of the will are formed, and from here all orders for the action of the voluntary muscles are issued.

Work of the Cerebellum. The work of the small brain seems to be to regulate the muscular movements which are directed to be made by the large brain. It has been discovered that when the cerebellum is injured, a person can not balance the body, as is required even in standing and much more in walking. A bird whose small brain is seriously injured or removed, can move its wings and its legs, but it can neither fly nor walk.

The Spinal Cord. The nervous matter of the brain is continued down through the back, passing through openings in the bones of the spinal column. This is called the *spinal cord.* Here (*131, last section of the head*) is where the spinal cord begins. This (124) upper part of the cord (*medulla oblongata*) is a very important part of the nervous system, for the reason that it seems to have control of some of the most vital operations of the body. When it is injured, the breathing muscles fail to act, which, of course, means instant death. Here (150) we see the spinal cord continued downward. Now we will turn again to the nervous plate of the Aid, where the whole of this great nervous cord is shown with its numerous branches of nerves.

The Nerves. The nerves are composed of the same substance as the brain. They are silvery threads which branch out from the brain and spinal cord and are distributed to all parts of the body. Twelve pairs pass out through openings of the cranium. These are called *cranial nerves.* Thirty-one pairs pass out from the spinal cord through openings of the back-bone, as shown on the plate. These are called *spinal nerves.* The cranial nerves go to the eye, ear, nose, tongue and other important organs. The spinal nerves go to the arms, trunk and legs.

Besides the nerves which branch out from the brain and spinal cord, there is, on each side of the back-bone, a chain of nerve centers—little bits of brains, as it were—running down through the body. From these small nerve knots, delicate nerves run out, some to the heart, lungs and stomach, and others' to the blood-vessels and to the cranial and spinal nerves. So all the important organs of the body are, in this way, connected with each other and with the brain. This figure (*The Sympathetic System*) shows, beautifully, this wonderful nervous connection. The interesting object of this arrangement—which is called the *sympathetic nervous system*—we shall soon learn.

Nervous Action. There are three kinds of nervous action. We will first consider the relation between mind, brain and nerve. The nervous system is very much like a telegraph system. The mind has been called the operator, the brain and spinal cord the sending or receiving offices or instruments, and the nerves the wires or lines running to all parts of the body. The comparison is very apt, indeed. One set of nerves runs *from* the brain or spinal cord to the muscles, so that every muscular fiber is in direct communication with head-quarters. Now, wherever a muscle is to act, every fiber of it, in some mysterious way, gets a message over its nerve line, from the nervous capital, directing it precisely how much to contract or relax. For example, you make up your mind to close your eyes. The order is sent out over the nerve lines which go to the fibers of the circular muscle which we have found to lie around the eye, and promptly the eyelids close. The nerves which carry messages to the muscles are called nerves of *motion.* Another set of nerves are called nerves of *feeling.* They carry impressions from the body *to* the brain. These nerves are distributed so thickly near the surface of the body, in the skin, that it would be almost impossible to find a point on the body where the **prick of**

a pin would not be felt. If you touch your body on its skin surface anywhere, even with the fine point of a needle, you are sure to disturb one or more of the nerves of feeling. Quicker than thought they report the impression, according to the degree of its severity, to the brain, which, if the situation at the surface demands it, will promptly return an order, over the nerves of motion, to the muscles of the endangered part, to do their best to get it out of the way of harm. For instance, a mosquito may alight on your forehead so lightly as to make no impression on your nerves of feeling, and, consequently, you are not aware of it. But now he punctures the skin and touches a nerve with his wonderful little stiletto. The news of his attack has been received by the brain, and an order sent back for defense and protection. Quicker than thought your hand has come up and routed or crushed the little assassin.

But the impressions which the nerves of feeling carry to the brain and mind are not all alarming or painful. Many of them are impressions of comfort or pleasure. A gentle breeze fans your body on a hot summer day. Hundreds of nerves are telling it to the mind, which enjoys it as a pleasure. Light impresses the nerve of sight, and beautiful views of form and color are spread before the mind. Sound excites the nerve of hearing and the charms of music are enjoyed. Invisible particles from a rose come in contact with the nerve of smell and we are delighted with the fragrance of the flower.

Reflex Nervous Action. If all the muscles were voluntary muscles, that is, if no movement of the organs of the body could be made without a special order from the mind, the continuance and enjoyment of life would be impossible. Every breath, every heart-beat, and many other operations of organs which can scarcely be dispensed with for even a few moments, would need to be constantly thought of and directed. Fortunately, the mind, and even the brain, is relieved from the ordinary control of the operations of organs upon whose regular and constant action our life depends. So the heart goes on beating, the lungs continue breathing and the stomach keeps on working, while the mind rests and the brain sleeps. Let us see how this is done.

The spinal cord may be regarded as a continuation of the brain. It is composed of the same two kinds of matter—white and gray. We may also look upon the spinal cord as a *deputy* brain. A deputy is appointed as a substitute for another, and empowered to act for him. An officer may have more duties to perform than he can personally attend to. So an assistant is given him, who is entrusted with certain lines of work for which he is held responsible. When serious questions or difficulties arise in the assistant's department of work, he appeals for special advice to the chief officer. So in the body, while the brain executes the orders of the mind, and controls the voluntary operations and movements of the body, the spinal cord is entrusted with the control of the involuntary muscles which perform the work of the heart, lungs, stomach and other vital organs, *except in cases of emergency.* For example, when food comes into the stomach, certain movements of the walls of that organ are necessary. So the food makes an impression on the nerves which report its presence, not to the brain or to the mind, as a sensation, but to the origin center of those nerves, in the spinal cord. Here the cord exercises its authority and returns (reflects) an order over motor nerves to the muscles of the stomach to perform the needed service.

In the same way, the presence of impure air or the absence of air in the lungs causes impressions which are carried to the cord, which returns orders for the action of the breathing-out or breathing-in muscles, as the case may be. All these performances go on steadily, whether we are awake or asleep. But when an emergency arises, as, for instance, if the muscles of the chest are strongly resisted in their efforts to expand it, by outside compression, the news of the trouble is carried beyond the nerve centers of the cord up to the brain, where the mind quickly grasps the situation and promptly issues orders for the best possible measures of relief. A familiar illustration of reflex action is found in the flapping of a fowl whose head has been cut off. Its muscles which produce its violent motions are not in connection with the brain, and can not be controlled by it. Each fall to the ground produces an impression which starts from the cord a message for the repetition of these muscular movements. Even when it seems to have settled down quietly, if you touch its body the movements will be renewed.

Sympathetic Nervous Action. We have seen how the sympathetic nerves connect important organs with each other and each with the brain. So if one organ suffers, the others suffer more or less with it. When the stomach is distressed, the head aches. When the heart's action is excited, the stomach is affected. When the brain is impressed with the mind's sense of shame or modesty, the little blood-vessels in the skin of the cheeks swell out and are more than usually filled with blood, and we call this delicate expression of their *sympathy*, blushing.

Health of the Nervous System. We would naturally suppose that organs so delicately constructed, and yet so prominent in the operations of the body as those of the nervous system, would need the most proper care to prevent their derangement or injury. And so it is. The brain needs especial care. It needs rest at proper intervals; not only from severe application, but the complete rest of *sleep*. An overworked brain is a diseased brain. On the other hand, the brain must have a proper amount of exercise to keep it in vigor. Besides healthy and varied exercise, the brain needs pure blood regularly and in proper quantities. Too much or too little blood will paralyze it. Hence its dependence on the proper action of the blood-*circulating* system. Impure blood will weaken its action. Hence its dependence on the blood-*purifying* system.

Severe excitement of the mind or long continued anxiety cripple the work of the brain, and finally result in insanity. A cheerful state of the mind is favorable to healthy nerves and long life. Consequently, all proper enjoyments, as the delights of music, pleasant changes of scenery, varied means of recreation and social pleasures, are like tonics to the nervous system first, and through it to the whole body.

There is no system of the body that is more severely outraged by the habit of drink and the use of narcotics than the nervous system. But this subject is so very important that it will be fully explained in a special chapter, after we are still better acquainted with the structure of the body.

OUTLINE.

THE NERVOUS SYSTEM.

WHAT?
- Consists of brain, spinal cord and nerves.
- Very soft and delicate in structure.
- Composed of white and gray matter.
- The large brain called the cerebrum.
- The smaller brain the cerebellum.
- Right and left halves of brain—called hemispheres.
- Nerves of two kinds, nerves of sensation or impression and nerves of motion.

WHERE?
- Brain enclosed in cranium.
- Spinal cord extends from base of brain through the spinal canal of back-bone.
- Nerves branch out from the brain, spinal cord and the sympathetic nerve knots, and are distributed to all parts of the body.

WHY?

- To serve the mind in directing the voluntary movements of the body.
- To control, by reflex action, the involuntary muscles.
- To bring to the mind, from the body and from the outside world, impressions producing the sensations or feeling of touch, taste, light, sound, smell, pain or pleasure.

QUESTIONS.

What are the organs of the nervous system?
What position in the body does the brain occupy?
Whose special instrument does the brain seem to be?
What can you say of the brain's structure?
What difference in the color of its substance?
Describe how the brain is protected.
What is the cerebrum?
What is the cerebellum?
What is the arbor vitæ?
What is meant by the hemispheres of the brain?
What can you tell of the work of the cerebrum?
What seems to be the use of the cerebellum?
Where is the spinal cord?
What is its upper part called?
What makes this part so very important?
What are the nerves?
From where do they start?
Where do they go?
How many pairs pass out from the skull?
What are these nerves called?
How many pairs branch off from the spinal cord?
What are these called?
Where do the cranial nerves chiefly go?
To what part are the spinal nerves chiefly sent?
What can you tell of the sympathetic nerves?
To what have we compared the nervous system?
Tell what you can of the comparison.
Do all the nerves perform the same kind of work?
Explain what is meant by the nerves of motion.

What is meant by the nerves of feeling?

Does the mind attend to all the movements of the body?

If not, will you explain your answer?

What kind of muscles are controlled by the nervous system independently of the mind?

What is such nervous action called?

Can you give an example of reflex action?

What is sympathetic nervous action?

Can you give an example of it?

Why does the nervous system need special care?

What kind of exercise is needed by the brain?

By what habits are these organs especially injured?

THE SPECIAL SENSES.

Nerves of Common Sensation. All the sensory nerves except four, are nerves of common sensation. They are distributed everywhere throughout the body. They need no special organs to enable them to receive impressions. Near the surface of the body or in the skin, they end in little folds or loops called *papillæ*. All the nerves of the sense of touch are nerves of common sensation.

Nerves of Special Sensation. There are four nerves of special sensation. These are the nerves of sight, hearing, smell and taste. Each of these nerves has a special organ without which no impression can be received to be carried to the brain. These organs are very delicate and wonderful structures. They are really special instruments of the nervous system.

THE EYE.

The *optic nerve*, or nerve of sight, is one of the nerves of special sensation. Starting from the brain, it passes to the eye to be distributed over the back, inner surface of the eyeball. This nerve is impressible only by light. But without the eye, the light would not impress it. The eye is an instrument to gather the light which is reflected to it from objects, and to bring it to bear on the optic nerve in such a way that an impression is made and carried to the brain, where the mind receives the impression as a picture of the object from which the light came. How all this is done, is very mysterious. But the organs which are concerned in the process can be easily examined and studied.

Protection of the Eye. We observe, first, that the eye is lodged in a deep socket of the bones of the head. Besides this feature of protection, there is placed behind and around the eye, quite a layer of fat, so that, even if the eye is struck, the force of the stroke is very much lessened by this fatty cushion. In front, it is guarded by the eyelids, eyebrows and eyelashes. The eyelids serve as a curtain. The eyebrows prevent the perspiration from running down from the forehead upon the lids. The eyelashes prevent dust from entering between the eyelids.

(Manikin of the Eye.)

Tear Apparatus. Turning aside this outer section which represents the natural open eye, we see a gland lying in the outer corner above the eye. This is called the *lachrymal* or *tear gland.* It secretes from the blood a watery fluid which it pours out upon the eyeball. By the act of winking, the eyeball is entirely bathed by this fluid, which, after it has flowed over the eye, collects in a little lake at the inner angle, from whence it is drained by two little channels (2) into the *tear duct* (1) which communicates with the nose. Shedding tears is simply an overflow of this eye-bathing fluid, when it is secreted in unusual quantity. At such times, the little channels cannot carry it away sufficiently rapid; so it flows over upon the cheeks. This unusual activity of the tear-gland may be produced by certain states of the mind, as sorrow or great joy; or by certain diseased conditions of the parts about the eye, as inflammation or a severe cold.

Oil-glands. In both the upper and lower eyelids numerous little glands are found (4) which secrete a kind of oil. From these little glands, very small ducts lead to the edges of the eyelids, where the oily matter is discharged through the little openings shown on the section (3). This oiling of the edges of the eyelids is very important. In the first place, the edges are thus prevented from sticking together. Further, were it not for this provision, the eye-bathing fluid would flow over the margin of the eyelids upon the cheeks. But since a watery liquid does not easily flow over an oily surface, such an overflow is prevented by the oiled edges, except when there is a flood of tears.

Eye Muscles. On the second eye-section a number of muscles which move the *eyelids* are shown. On the third section four of the six muscles which move the eyeball are represented (9, 10, 11, 12). These are called the straight muscles of the eye, because each of them draws or rolls the eye in the direction of its contraction—upward, downward, inward or outward. There are also two other eye-rolling muscles. They are called the oblique muscles of the eye, and give it its peculiar *rolling* motions.

The White Coat of the Eyeball. The eyeball has three coats. The outer coat, or white of the eye, is called the *sclerotic coat* (15). It is a strong, tough membrane which forms quite a substantial case into which the *cornea* (14) is set in front, like the glass or crystal of a watch. The sclerotic coat is not sensitive; that is, it has no nerves of feeling. But it is covered, in front, with a very delicate membrane which contains very fine blood vessels (13) and nerves. When these little vessels become swollen with an unusual amount of blood, the eye is said to be "blood-shot," and when a cinder or dust grain lodges on the eye and makes an impression on the delicate nerves of this fine protecting veil, the sensation is very painful. No light passes through the sclerotic coat; but the cornea—the window of the eye—is very transparent.

The Black Coat of the Eye. Next to the tough, outside white coat lies the *choroid* (*fourth section*, 18). This is a soft black membrane. It prevents

the reflection of strong light from the inner surface
of the eyeball, and thus serves an important part in
making the sight sharp and clear. The front part
of the choroid coat is arranged like a circular curtain. This is called the *iris* (20). This is what gives
the eye its so-called color. The difference between a
black eye and a blue eye is, that the cells of the iris of
the one have a black coloring matter in them, while the
cells of the iris of the other contain blue coloring
matter. In the center of the iris is a circular opening
called the *pupil* (21), which you can see by looking directly into the eye of another person who stands close
before you. Through this little circular opening,
surrounded by the curtains of the iris, the light must
pass on its way to the back inner part of the eyeball.
The amount of light which passes through the pupil is
regulated by an interesting action of the iris. When
the light is strong, the little muscles which are
threaded through the iris-curtain in two sets (a, b),
produce the effect of making the pupil smaller, so
as to pass less light. When we go from a light
place into the dark, these same muscles bring about
an opposite effect, that is, the pupil is made larger,
so as to admit more rays. This adjustment of the
curtain of the eye is not instantly done. It requires
some time. This you can easily observe; for in
going from the dark into a very light room, you can
not see well until the change in the size of the pupil
of your eye has been made. So, also, when going from
a bright room into a dark place, at first it seems to

be "pitch dark;" but, by and by, when your eye is adjusted to the change, you will be surprised to find that it is not so dark after all.

(*Sixth Eye Section.*)

The Nervous Coat. The third or inner coat of the eye is the *retina* (27). This lies only over the back inner part of the eye-ball. It is really the end of the optic nerve, or nerve of sight, spread out to receive the impression of the light in the eye. The retina is an exceedingly delicate nervous screen on which the action of the different parts of the eye makes a picture of the object we look at. How this picture is carried by the nerve to the brain, and there grasped or perceived by the mind, we do not understand.

The eye section last referred to shows also how numerous blood-vessels are distributed over the inner part of the eye. These enter the eyeball from behind with the optic nerve as shown at (25). This place where these blood-vessels and nerve enter is called the "blind spot," because the light rays which fall there make no impression on the nerve. The most sensitive place on the retina is at a point at a little distance from the "blind spot." This is called the "yellow spot" (24). When the eye is perfectly adjusted to perceive an object, the image will fall on the retina at or near this spot.

The seventh section gives us another view of the dark choroid coat (28) as it lies behind the retina

(27), and also the opening through the choroid (29) through which the optic nerve enters. The last section shows veins and arteries (24, 25) distributed through the dark coat. A portion of the choroid is represented as being removed to show a small part of the sclerotic coat behind and outside of it, and also the entrance of the optic nerve, with its accompanying blood-vessels (23) through the outer white coat.

The Humors of the Eye. Between the cornea and the iris, in the front part of the eye, is a watery fluid called the *aqueous humor* (*see 18, sectional figure of eye, fifth chart*). Back of the iris lies the *crystalline lens* (23). This is a beautiful gem-like little body, as clear as a crystal. Back of the lens, and between it and the retina, the eyeball is filled with another clear jelly-like substance called the *vitreous humor* (25). The effect of these three humors which are contained in the eye, and especially the effect of the crystalline lens, is to produce the image of things on the retina, as has already been mentioned.

THE EAR.

The Nerve of Hearing. The nerve of special sensation which goes from the brain to the ear is called the *auditory nerve*. As the nerve of sight is sensitive only to light, this is sensitive only to sound. The ear is an instrument to collect sounds and bring them to bear on the auditory nerve in

such a way that an impression is made and carried
to the brain to be recognized by the mind.

Outer Ear. The ear is divided into the *outer, middle*
and *inner* ear. The outer ear has a
more or less cartilaginous frame. This
allows motion, and, at the same time, keeps it in
shape and position. It has also a few small muscles.
But in the human ear these are nearly altogether
useless, since men do not move or flop their ears.
In animals which move their ears in various ways,
these muscles are quite well developed.

(*Ear Figure, Fifth Chart.*)

The Middle Ear. From the outer ear (1) a tube, a little
over an inch long, called the *auditory
canal* (2), leads in to the middle ear,
where it is closed by a membrane called the *mem-
brana tympani,* which means the membrane of the
tympanum or drum (3). The middle ear is often
called the "ear-drum," and the membrane just mentioned may be called the "drum-head," for it does,
indeed, act very much like the head of a drum.
Between this membrane of the drum and its inner,
opposite side or end, there is stretched a very curious
little suspension bridge of four small bones. The
first of these is attached to the drum-head, and from
its shape like a hammer is called the *malleus* (4).
The next is called the *incus,* because it is shaped
like an anvil (7). The third is a very small pebble
of a bone called the *orbicular* or round bone (10). It

is followed by the *stapes* or stirrup bone (12), the last span in the little bridge. This rests against a small window-like membrane which is stretched over an opening in the inner side of the drum. At the bottom of the ear-drum or middle ear is an opening into a tube which leads from the ear to the throat. This is called the *Eustachian tube.* Its object is to supply the ear-drum with air, for without air inside to balance the pressure of the air on the outside of the drum-head, the action of the latter would be very imperfect and our hearing, in consequence, very dull. We frequently experience the truth of this statement; for whenever the *Eustachian tube* becomes clogged, as in the case of a very bad cold, our hearing is very much impaired.

The Inner Ear. The inner ear is carefully hidden in a hollow place in the solid bone. In that part of it which lies next to the middle ear, is a little hall-way, or *vestibule*, about as large as a grain of wheat. This leads, on one side, into the arched or semi-circular hall-ways which are called the *semi-circular canals* (13, 14, 15). On the other side the vestibule opens into the *cochlea* (16, 17), which is shaped like a snail shell or a tiny winding stair. Here the auditory nerve, or nerve of hearing, takes up the impression of a sound and transmits it to the brain.

How We Hear. All sounds are produced by the vibrations of bodies. To make this plainer, when a bell is struck its particles are

thrown into a violent trembling. By these trem-
blings or vibrations of the material of the bell the
air is thrown into a wave-like motion all around it.
When these trembling air waves reach the ear, the
sensation of sound is produced and we say we hear
the bell. When a person speaks to us the voice
chords in his throat are set into rapid vibration.
These vibrations produce waves of sound in the air;
the air carries these waves to the ear, where, passing
in through the auditory canal they tremblingly beat
upon the drum-head; this carries the sounds to the
bridge of little bones. Having passed over these,
it enters the vestibule, then vibrates into the semi-
circular canals and rebounds into the cochlea, where,
as already stated, it is taken up by the nerve and
carried to the brain where the mind interprets it as
the voice and the language of the speaker.

THE SENSE OF SMELL.

The organ of smell is the nose and its cavities.
The nerve of smell is called the *olfactory nerve.*
This nerve is spread out in many branches over the
delicate mucous membrane which lines the inside of
the nose. To make the surface on which the nerve
of smell is distributed as large as possible, there is
set into the nostrils, against the outer walls, a pair
of scroll-like bones. These are the *turbinated bones*
of the face. Over their winding surfaces, covered
with the mucous lining, the nerve of smell is spread.
The two small nasal bones unite the nose to the

skull and keep it in shape. The lower part of the nose is shaped by a frame of cartilage, the advantage of which over a nose-frame of solid bone you can readily see.

How We Perceive Odors. Things which have an odor, or smell, give out little particles of matter, altogether invisible. As these float in the air, they are drawn into the breathing passages of the nose and mouth at every breath. Of course those which pass with the air into the mouth can make no impression of smell, for there are no nerves there which are affected by odors. But those which pass into the nostrils strike upon the olfactory nerve branches which, as we have seen, have their special location there. The mind, receiving these impressions, recognizes the odor, which may be feeble or strong, agreeable or very unpleasant.

The sense of smell affords us protection in two important ways. Its organ, the nose, is set at the very gates of entrance of the air we breathe and the food we eat. So when the air is filled with putrid or offensive invisible matter, which, of course, would make it unfit to breathe, we are cautioned by the sense of smell, and instinctively turn away and seek a purer air to breathe. Likewise, we are often warned, just in time, against putting into the mouth, as food, substances whose odor betrays their unfitness to be eaten. Fortunately it is so provided in nature that poisonous and other harmful substances have generally a strong and peculiar smell, although this is by no means always the case.

THE SENSE OF TASTE.

The Tongue. The special nerves of taste have their loop-like endings chiefly in the tongue, which is, consequently, usually spoken of as the organ of taste. But these papillæ, or end expansions of the nerve of taste are also distributed over the walls of the back part of the mouth. On account of the numerous little folds of nerve endings on the tongue, this organ has quite a velvety appearance. Besides serving as the chief organ of the sense of taste, the tongue also aids in the chewing of the food and in producing the sounds of speech.

How We Perceive Taste. When substances which have a taste come in contact with the papillæ or nerve loops of the tongue, the impression is at once carried to the brain and mind. In order that such an impression can be made the substance to be tasted must be in a dissolved state. No dry or solid substance can be tasted. So the mouth is kept moist, and, as we shall learn later, during the process of eating, a large quantity of saliva is thrown into the mouth. This dissolves at least a portion of the food or substance which is in the mouth, so that its taste is well perceived. When the mouth is dry from disease, or from great thirst, food has but little taste and is very unpalatable. So when the nerves of the tongue are covered with a strange coat, as in disease, our food does not taste natural.

OUTLINE.

ORGANS OF THE SPECIAL SENSES.

WHAT ?
- Eye—the organ of sight.
- Ear—the organ of hearing.
- Nose—the organ of smell.
- Tongue—the organ of taste.

WHERE?
- Eye—under arch of frontal bone.
- Ear—in hollow of temporal bone.
- Nose—at entrance of air and food passages.
- Tongue—lies on the floor of the mouth.

WHY ?
- Eye—to collect rays of light from objects and produce a picture or image of such objects on the expansion of the nerve of sight.
- Ear—to collect sound waves and convey them to the nerve of hearing.
- Nose—to bring odorous matter in contact with the nerve of smell.
- Tongue—to bring substances having taste in contact with the nerve of taste.

QUESTIONS.

Where are the nerves of common sensation distributed ?
How many nerves of special sensation are there ?
Can these nerves receive impressions directly ?
What is the special instrument of the nerve of sight ?
What is the proper name of the nerve of sight ?
By what only is it impressible ?
What is the use of the eye ?
How is the eye protected by its position ?
What other means of protection are furnished it ?
What is the use of the lachrymal or tear gland ?
Where is this gland situated ?
After bathing the eye, how is that fluid drained away ?
What is meant by "shedding tears" ?
What conditions of mind and body may cause this

How many coats has the eyeball ?
What is the nature of the outer coat ?
What is the cornea ?
Is the white coat of the eye sensitive ?
How do you account for the pain felt when a cinder lodges
    on the eye ?
What is meant by the eye being "bloodshot" ?
Does light enter the eye through the white coat ?
Through what does it enter ?
What is the color of the middle or choroid coat ?
What purpose does it serve in the eyeball ?
Where and what is the iris ?
What gives the eye its color ?
What is the pupil ?
How is the pupil regulated to admit more or less light ?
Where and what is the retina ?
What is formed by the eye on the screen of the retina ?
What humor lies between the cornea and iris ?
Where and what is the crystalline lens ?
What humor occupies the back part of the eye ?
Which of these parts is most effective in collecting the light
    on the retina ?

What is the nerve of hearing called ?
By what only is it impressible ?
What is the work of the ear ?
Into what parts is the ear divided ?
What tube leads from the outer ear to the "drum" ?
Describe the ear drum.
What is the use of the Eustachian tube ?
What are the parts of the inner ear ?
By what are all sounds produced ?
Describe the course of sound-waves from a sounding body
    through the ear to the nerve of hearing.

What is the organ of the sense of smell ?
What is the name of the nerve of smell ?
To what is this nerve sensitive ?
Against what does the nerve of smell afford us protection.

Where are the extremities of the nerve of taste located ?

What gives the tongue its velvety appearance ?

Why are dry or solid substances tasteless ?

What provision is made to make food more perceptible to the taste ?

Why does food have no taste to us when we are sick ?

THE CIRCULATORY SYSTEM.

Body-Building. The building of the body, that is, its growth to full stature, and to the complete development of all its parts — bones, muscles, and internal organs — continues during the first twenty or twenty-five years of life. As in the building of a house, there must be brought to every part of the rising structure, the precise kind of material then and there needed, so in the building of the body, there must be delivered at the proper times, and in proper places, all the various kinds of material needed for the perfect development of the physical structure of the body according to the beautiful design and the wise plan of its Divine Architect.

As the building of a house requires many kinds of things, such as wood, stone, iron, glass, lime, sand and putty, so the structure of the body calls for material suitable for bone, muscle, nerve, hair, nails, and so on. Not only must such body-building material be supplied in proper kind and quantity, but it must be furnished with the utmost promptness and regularity.   It is an interesting fact that while Nature will not allow herself to be hurried in the work of rearing a human body, but, instead, carries on the process deliberately through a score or more of years, she will, on the other hand, be unable to do her work well, and at the end of twenty or twenty-five years, present one of her master-pieces, in the form of a perfectly developed human frame, if she is hindered in her work from a lack of proper and sufficient building material.   The result of body growth under such circumstances is general weakness, deficiency in size and deformity of structure.   On the other hand, if during the "growing years" of life, there is a steady supply of building material furnished throughout the body, and Nature is permitted to lay up such material in proper kind and proper quantity in its appropriate places, without hindrance or interruption, the result will be a human form, measuring up to the full strength and stature of perfect physical manhood and womanhood.

Repair.   Further, besides the material which the body needs for its growth, it is constantly exposed to wear and tear, and, consequently, it must be supplied continually with material for repair.   It is impossible altogether to avoid the wearing out of parts of the body.   Some of the muscles, like those of the heart, for example, are on constant duty.   Every contraction of a muscle destroys a part of its fiber. The nervous system is also constantly suffering wear. The slightest effort of body or mind produces damage which must be made good to maintain our strength. The simplest thought which occupies the mind lays a tax on the structure of the brain. Either a wink or a whisper destroys muscular fibers. If even these gentle movements are wearing, how great must be the destruction throughout the body, by labor which exercises vigorously the brain and many muscles.

So a system of organs is provided whose work it is to carry to all parts a supply of material as may be needed for building or repairing. This is the *circulatory system.* The blood — a bright red fluid with which we are all familiar — floats the building and repairing material through the channels and reservoirs of the circulatory system. Let us learn some more about this important fluid.

Blood. Strange as it may seem to you, the liquid of the blood is no more red than water is red. The blood is really a thin, watery fluid in which millions of little bodies called *corpuscles* float. The colorless liquid in which these corpuscles float is called the *plasma.* Looking at blood with the naked eye we do not even suspect its interesting composition. But the microscope reveals wonders in every drop of this life-giving fluid. It shows that the corpuscles are of two kinds — red and white. The red corpuscles are by far the most numerous, and give the blood its bright red color. They have a curious tendency to arrange themselves like a pile of coins. Both the red and the white corpuscles, as they appear under the microscope, are shown on the sixth plate of the Aid.

Organs of the Circulatory System. In order that the blood may be carried to even the remotest parts of the body, there must be larger and smaller vessels to serve as channels through which it may flow in streams of different size. It requires also a propelling organ which shall cause the blood to flow steadily and in sufficient quantity through the circulation channels. All these organs and vessels are properly provided in the body. The *heart* is the propelling engine. The arteries, veins and capillaries are the circulatory channels.

The Heart. (*Refer to fourth chart.*) The heart lies near the center of the chest, a little to the left of the middle line. A man's heart is about as large as his fist. It is a very strong muscular pump or engine, and does an enormous amount of very important work, as we shall soon see. It has four reservoirs or chambers — two on each side. The upper chamber of the right side of the heart is called the *right auricle* (U); below this (W) is the *right ventricle*. The *left auricle* (V) forms the upper chamber, and the *left ventricle* (X) the lower chamber of the left side of the heart. Between the chambers of the right side and those of the left side is a closed partition. So the heart may be regarded as a double organ. The upper and lower chambers — that is, the auricles and ventricles — are separated by valves, whose important use we shall learn when we trace the course of the blood through the heart.

Covering of the Heart. The heart is covered with a loose sac which is very delicate and as smooth as satin. This covering sac has been called the heart's "night-cap"—a name which would be much more appropriate were it not for the fact that the heart wears this "cap" both night and day. Physiologists call it the *pericardium*, which means, about the heart. The pericardium secretes from the blood a thin fluid which gathers on the inner side of the sac. In this way the surface of the heart is kept smooth and moist, so that all harm from friction, on account of the heart's movements, is prevented. This is another illustration of the divine foresight and care in planning the wonderful structure of our bodies.

Arteries and Veins. The chart gives us a good general view of the arterial and venous blood-channels of the body. Having observed carefully the *position* of the heart, notice the distribution of arteries and veins throughout every part. With one exception (the pulmonary artery, 16), the arteries are shown in red. The veins, with one exception (the pulmonary veins, 31), are shown in blue. The same exceptions being made, the arteries carry the pure blood and the veins the impure. In the general view we are now taking of the circulation, as shown on the chart, we observe the great artery called the *aorta* (15) starting from the heart. It sends great branches upward (19, 21) to the neck and head, and sideward (20, 22) through shoulders, arms, hands and fingers. The aorta, after making a great bend or *arch* (17), passes downward through the trunk of the body. As it descends, it gives off important branches to the great internal organs as well as to the walls of the chest and abdomen. In the lower part of the abdomen, two great arterial branches are formed (42) to be again subdivided and distributed to the hip region and down through the lower extremities even to the ends of the toes.

While the left side of the figure shows the distribution of arteries, the right side is made to show the venous system. We see the veins start from all parts as tiny vessels, collecting into larger channels, and finally entering from below into the great ascending vein (47) and from above into the great descending vein (14), both of which empty into the right upper chamber of the heart, that is, into the right auricle (U).

Intersections. We find a wise provision in the cross-connections which are found in the circulatory channels in many parts of the body. Both the arteries and the veins have such connecting branches, especially at places where an obstruction to the free flow of the blood through any one channel is liable to occur. This arrangement is well shown on the chart in the veins of the right leg.

Structure of the Arteries. The arteries are very strong elastic tubes, through which the blood is driven with much force by the strong propelling action of the heart. Since the wounding of an artery is liable to result in a loss of blood, to a dangerous extent, the arteries are, as a rule, deeply imbedded among the muscles and other tissues of the body. At a few places where they lie quite near to the surface, as at the temples, or at the wrist, the throbbing of the blood as it is driven through the artery by the action of the heart, may be plainly felt. These throbbings are called the *pulse.* The interior structure and the various coats of an artery, as they appear under the microscope, are shown on the sixth chart.

Structure of the Veins. The veins are much more limp, or less elastic than the arteries, and the blood flows through them at a much more sluggish rate. As a rule, the veins lie much nearer to the surface of the body than the arteries. They may frequently be easily traced as the bluish blood flows through them close beneath the skin. To prevent the blood in the veins from flowing in a backward direction, they are provided with valves. These valves, together with the general structure of a vein, are also represented, in enlarged view, on the sixth chart.

The *capillaries* are very fine tubes con-
The Capillaries. necting the arteries with the veins.
Their name comes from a Latin word
which means, *a hair*. It is difficult to imagine how
numerous these little capillary blood-vessels are, and
how well they are distributed to all parts of the body.
You could scarcely prick your skin with a needle anywhere without bringing some blood to the surface;
you are sure to pierce some capillary and cause it
to leak.

We have now described the heart and
The Course of the Blood. the blood-vessels of the circulatory
system. Let us next trace and learn
the course of the blood through these organs. (*Refer to sixth chart*). This figure in the middle of this
chart will help us clearly to understand it. Here the
heart is laid open, showing its inner chambers and
the valves between them. This large blood-vessel
(27) is one of the two great veins which empty the
impure blood from the body into the right auricle of
the heart. It is called the *descending vena cava*,
because it brings the blood from parts above the
heart. Here (51) is the other of these large veins
—the *ascending vena cava*, bringing the impure blood
from the lower parts. From the right auricle (28)
the blood goes through a valve-like partition into the
right ventricle (29). The object of the valve is to
prevent the blood from going back from the right
ventricle into the auricle above, which would otherwise surely happen when the muscles of the ventricle

contract to drive the blood forward in its course.
From the right ventricle the blood goes out of the
heart to the lungs through the *pulmonary artery*
(32). This great artery, as we see, is divided into
branches whose numerous sub-divisions are distributed through the right and left lungs, where the
process of purifying the blood is performed, in a
manner which will be described in the next chapter.
Coming back from the lungs to the heart, the blood
flows through the *pulmonary vein* (33), which
empties into the left auricle. From here it passes
through a valve into the left ventricle. Now notice
the thick muscle of the left ventricle. When this
strong muscle contracts, the blood is forced out
through the *aorta* (34), which branches out into
many arteries, and then into capillaries, all through
the body, as is here beautifully shown. From the
capillaries the blood is collected by the smallest veinlets, to be carried into larger streams, which farther
unite to bring it into the great ascending and descending veins, with which we began our tracing of its
course. Wonderful to tell, this whole circuit is made
in less than half a minute!

It is well for you to learn and trace the
Summarized Course. course of the blood in this way: Coming impure from the body it flows into
the right auricle; then through the valve into the
right ventricle; then through the pulmonary artery
to the lungs; then through the pulmonary veins to
the left auricle; then through the valve into the left

ventricle; then through the aorta into many arteries; then into the capillaries; then into the veins, which return it to the heart.

Three Divisions of the Circulation. There are really three divisions of the circulatory system. The first is the course of the blood from the heart through the body *for building and repair.* This is called the *systemic circulation.* The second is its course from the heart through the lungs *for its purification.* This is called the *pulmonary circulation.* The third is a special course of a part of the blood through the liver. This is called the *portal circulation.*

The Portal Division. The portal circulation is a special feature of the systemic division. We see here (58, 59) how several of these prominent veins gather the impure blood from these lower organs in the body, and, gathering into this large *portal vein* (60), it is thrown into the liver. As we shall learn further on, this blood also contains food substances which the veins have absorbed from the stomach and intestines. After being partly purified by the action of the liver, the blood is again collected into the *hepatic vein* (52) which turns it, as you see, into the large ascending vein (51) which goes to the heart.

Blood change in the Capillaries. In the capillaries are the landing places where the little cargoes of building material, which have been floated from the heart through the arterial rivers, are unloaded and distributed to the thousands of little working cells which are everywhere busy in building or repairing the body. At some places material for muscle is unloaded; at others material for bone, nerve, or finger-nail is wanted. In exchange for this new material which the capillaries distribute *to* the body, they take back *from* the body the material which has become old, worn-out and unfit for use. The consequence is that the blood which has come from the arteries into the capillaries red and pure, leaves them and gathers in the veins, dark and impure. It would be altogether unfit to make another round through the body without being purified; so the capillaries deliver it to the veins, and these carry it to the heart, which drives it to the lungs—organs of another system—where its worn-out matter is unloaded, and it is again made fit to feed the body.

The Muscular Work of the Heart. The heart is, by far, the strongest muscular part of the body. No engine in the world, of its size, has so much strength. By the contraction and relaxation of its muscular walls it receives the blood into its reservoirs and again propels it onward in its course through the body. When the auricle muscles relax the ventricle walls contract. The result is that blood flows *into* the auricles—on the right side from the body, on the left side from the lungs—and *out of* the ventricles—on the right side towards the lungs, on the left side towards the body. When, next, the auricle walls contract and the ventricle muscles relax, the blood flows from the auricles into the ventricles. This wonderful "rhythm of life" is kept up from birth to death—sometimes through even more than "three score years and ten"—at the rate of 100,000 beats per day, 40,000,000 beats per year, and in the life-time of an octogenarian, 3,000,000,000 times without a stop!

Blood-vessel Wounds. The rapidity with which the blood escapes from a wounded blood-vessel depends upon the severity of the injury and the kind and size of the wounded vessel. From a pierced capillary, the blood may leak only in tiny drops. From a wounded vein the blood may flow in considerable quantity, but its escape from a vein is usually steady, and, in most cases, soon stopped by Nature's method of forming a "clot" in the wound. But from a wounded artery the blood "jets" with much force, to a considerable distance, and often in dangerous quantity. It is easy to tell whether a vein or an artery has been cut, by the greater force with which the blood spurts from the latter.

Practical Suggestions. Usually Nature can help herself in stopping the bleeding of a wound, so that in most cases no artificial measures need be taken to stop the flow. However, cases do frequently occur where the blood escapes from a wounded artery with so much force as to prevent the formation of a clot to stop the bleeding. Hence, bleeding to death has often occurred. In case of a serious cut of an artery on one of the limbs, a handkerchief or other bandage should be promptly bound about the limb *above* the wound, and drawn as tight as possible. It may even be necessary to twist the bandage down, using a stick as a lever, until the flow of blood through the injured artery is stopped by the pressure of the twisted knot.

Health of the Circulatory System. The steady and thorough work of the organs of the circulatory system is very essential to life and health. Any cause which tends seriously to increase or diminish the normal rate of the heart's action, is a thing to be avoided. The ordinary quickening of the flow of the blood, as in moderate exercise of the body, is not only harmless, but healthful. Here, again, *exercise* must be commended as a prime condition of a healthy circulation. Any part of the body which, from any cause whatever, remains comparatively unused, will not be supplied by the circulatory system with a sufficient quantity of pure blood to maintain its vigor. Such a part will therefore gradually wither and die. It follows that we need that kind of exercise regularly which will call into use all parts of the body, and thus prompt the flow of blood into every nook and corner of our physical structure.

OUTLINE.

THE CIRCULATORY SYSTEM.

WHAT?
{ Central organ, the heart.
{ Arteries, blood-vessels leading *from* the heart.
{ Veins, blood-vessels leading *to* the heart.

WHAT?
- Capillaries, uniting arteries and veins.
- Three divisions of the circulation:
- From heart to lungs and back to heart, for purification—*pulmonary circulation.*
- From heart to body and back to heart, for nutrition—*systemic circulation.*
- From veins of digestive organs through liver, for partial purification—*portal circulation.*

WHERE?
- Heart in chest, near middle.
- Arteries, veins and capillaries distributed throughout the body.

WHY?
- To carry pure blood to all parts of the body.
- To gather up useless or waste material and carry it to the organs which remove it from the system.

QUESTIONS.

What can you say of the growth or building of the body?
What can you say of the "wear and tear" of the body?
What is the object of the circulatory system?
What is the use of the blood?
What are the organs and vessels of the circulation?
Where is the heart situated?
About how large is the human heart?
Tell what you can about its purpose.
How many chambers has the heart?
Where is the right auricle?
Where is the right ventricle?
Where is the left auricle?
Where is the left ventricle?
What are the arteries?
What kind of blood do the arteries usually carry?
What are the veins?
Do they usually carry pure or impure blood?
What kind of blood flows through the pulmonary artery?
What kind of blood flows through the pulmonary vein?
Why do the veins have valves?
What are the capillaries?
What change in the blood takes place in the capillaries?

By what large vein is the blood from the lower body brought to the heart?

By what large vein is the blood from the upper body brought to the heart?

What prevents the blood from flowing backward from the ventricle to the auricle?

What is the work of the pulmonary artery?

What of the pulmonary veins?

Why is the muscle of the left heart extra strong?

What is the aorta?

Now trace the course of the blood, beginning where it comes impure to the right auricle.

How many divisions of the circulatory system?

Why does the heart send the blood to the lungs?

Why through all parts of the body?

Why is a large part of the blood carried through the liver?

What is meant by the pulse?

What is the effect of exercise on the circulation?

May exercise become too violent?

How is the circulation affected in an unused part of the body?

What is the best kind of exercise?

NOTE.—Teachers as well as pupils who may desire a fuller description of the distribution of the arterial and venous systems, will find it in the supplementary part of this book.

THE RESPIRATORY SYSTEM.

We have learned that the blood, in every round of its circulation through the body, is made quite impure by the worn-out or waste material which it takes on in the capillaries, and therefore must undergo a purifying process to make it fit to repeat its course through the system. For this purifying purpose the body is furnished with the *respiratory system.*

Work of the Respiratory System. The work assigned to this system is twofold in its nature. It must supply the blood, at every round, with a new supply of oxygen, and, at the same time, unload from the blood the impure matter which it brings to the place of purification. The chief organs of this system by which this double work is accomplished, are the *lungs*, which lie in the chest, close around the heart, the two organs together completely filling the cavity of the chest. Let us get a clear and accurate idea where these important organs are situated. (*Turn to the body manikin.*) Here is a manikin of the body. Removing first the skin, then the outer muscles of the trunk, we have the ribs (3, 4, 5) before us. Removing these, the contents of the chest are shown precisely in their natural places. These (8, 9) are the lungs. The heart lies immediately under and between them. Below the heart and the lungs, that is, between the chest and the abdomen, lies this strong, flat muscle (26) called the *diaphragm*, whose important service in the work of the respiratory system we shall learn about farther on.

Structure of the Lungs. The lungs are very spongy and light, being composed largely of air-cells, whose walls are very delicate. The right lung has three lobes (a, b, c) and the left lung two (d, e). The air-cells of the lungs, are all connected with tubes called *bronchial tubes* (25, *third section of lungs*). An inflammation of these tubes is called *bronchitis*. Besides the air-passages and air-cells which make up a great part of the bulk of the lungs, they contain numerous larger and smaller blood-vessels which carry the blood into them and *through* them from the heart, and *out* of them back to the heart.

The Trachea. The bronchial tubes unite in one large air-passage (24) called the *trachea* or wind-pipe. In the upper part of the wind-pipe, which comes close up to the mouth, the instrument of the voice called the *larynx*, is situated. It is this enlarged portion of the trachea (21, 22, 23). The more definite structure of the larynx or voice-organ is shown by the special manikin overlying the parts just referred to. It will be spoken of more particularly hereafter. The arrangement of the air-cells and air-passages reminds one of an inverted tree. The larynx, or voice-organ, corresponds to the lower and thicker part of the trunk of the tree; the trachea to the trunk itself; the branching air-tubes in the lungs to the branches and twigs, and the air-cells to the leaves.

What takes place in the Lungs? The pulmonary artery, which brings the impure blood from the heart into the lungs (16), branches out into many smaller arteries and still farther into a great many capillary tubes which wind among and around the numerous air-cells. When the cells of the lungs are filled with air, the oxygen of the air, in a wonderful way, passes through the wall of the cell and the wall of the capillary and unites with the blood. At the same time, impurities from the blood pass through capillary wall and cell-wall into the cells and out through the air-passages with the escaping breath. In this way the blood is renewed, purified and brightened by the life-giving oxygen, and starts off vigorously on another round through the body. The air-tubes, cells and capillary net work in the lungs, and the decided change in the color of the blood after being relieved of its impurities and taking on a fresh supply of oxygen, are nicely shown in the figure on the right hand upper corner of the sixth chart.

Breathing. Breathing is the act of the body by which the lungs are filled with air and emptied again at proper intervals. This operation is so important and essential to life, that it has been entrusted to the performance of muscles of the involuntary kind, that is, such as are not dependent on the direction of the mind. It is true, the will may interfere with the work of these muscles, so that we may suspend breathing to some extent. But it is Nature's plan that this work should be committed to faithful nerve-centers and muscles appointed for the purpose. It is well that this is so, for, otherwise, during sleep or other unconscious moments, when the mind gives no direction to the body, breathing would stop and life would end. Even when awake and in health, in this very busy age, we might *forget* to breathe.

How we Breathe. The scientific name for breathing is *respiration.* The act of respiration which brings air *into* the lungs is called *inspiration;* that which drives the air *from* the lungs, *expiration.* There are quite a number of muscles concerned in these acts. When the lungs are to be filled, these muscles expand the walls of the chest so as to enlarge the space inside. The air rushes in through the mouth, nose and wind-pipe, and fills the cells. Then a reverse action is produced by the muscles. The chest contracts and the air is forced out from the lungs by the same way through which it entered, but robbed of its oxygen and mixed with gases and impurities discharged from the body.

Chief Breathing Muscles. Between the chest and abdomen is the broad partition muscle, the *diaphragm* (26) which has already been mentioned. This muscle is chiefly concerned in ordinary, gentle breathing. When the lungs are to be filled, the diaphragm moves downward, pressing upon the contents of the abdomen. At the same time the muscles of the abdomen relax to make more room for the abdominal organs which are pressed down by the descending diaphragm. The outer motion of the abdomen wall can be seen and felt at every inspiration. Thus the cavity of the chest is enlarged, giving the lungs room for full expansion, provided the act of inspiration is unrestricted and complete. In expiration, the diaphragm rises and diminishes the cavity of the chest, forcing the air out of the lungs.

Besides the movements of the diaphragm in alternately enlarging and diminishing the capacity of the chest, there are other muscles concerned in the act of breathing. This is especially true in *forced* breathing. Referring to the rib section of the manikin, you observe muscles extending in different directions between neighboring ribs (6). These are called the *intercostal muscles.* On account of their various directions, their contraction moves the ribs, in various ways producing as many alterations of the size of the chest cavity. Likewise, there are other muscles which take more or less part in breathing.

Pleura. The inside of the chest is lined with a delicate web called the *pleura.* On the rib section the left side is shown with the intercostal muscles removed, exposing the pleura (7) within. This membrane is also spread as a covering over the lungs. It secretes a watery fluid which keeps the walls of the chest and the surface of the lungs moist, and thus prevents friction which would otherwise be produced in the movements of breathing. An inflammation of the pleura is called *pleurisy.* When both pleura and lungs are inflamed it is called *pleura-pneumonia.*

Health of the Respiratory Organs. From what we have learned of the structure of the chest and the action of the lungs in breathing, we cannot fail to see that the healthy and natural action of the respiratory system requires perfect freedom of motion or expansion of every part concerned in the vital act of respiration. It is Nature's plan that every air-cell in the lungs should perform its appointed part with all the rest, at every breath. If the habit of breathing full and deep, when the chest is perfectly free to expand and the body is in erect or straight position, is well formed, every lung-cell will be filled at each inspiration. But if the chest is in any degree restricted and compressed, the lungs will be but partly filled, and many of the air-cells will lose their elasticity, and finally become utterly useless. Such an injurious interference with the natural expansion of the chest may be produced by habitual unnatural positions in sitting or walking, or by wearing the clothing too tight about the body.

Exercise. We have learned that the flow of blood through the vessels of the circulatory system is much quickened by exercise. So the rapidity of the flow of the blood through the lungs depends very much on the degree of our bodily activity. When we lie in bed, for instance, the circulation goes on very steadily, and our breathing is performed very moderately and quietly. But as soon as we arise and move about, both the circulation and respiration are quickened. The more vigorous the activity of the body, the more air is drawn into the lungs to purify the greater quantity of flowing blood. The chest muscles act more strongly, and every cell in the lungs is inflated. All this tends to produce pure blood and active lungs, and, consequently, good health.

The *quality* or *purity* of the air which we
Impurities of the Breath. breathe is quite as important as the *quantity* which we inhale.   We have learned
that the oxygen, which is one of the gases or ele-
ments which compose the air, is taken from the lung-
cells to unite with the blood.   This alone would
make the breath which is given out impure because
of its having been robbed of its oxygen.   But the
air which is forced out from the lungs is made much
more impure by the gases which come from the
impure blood of the body.   These gases escape from
the lungs at every breath.   It is plain that if we
breathe in a close room, or in a confined body of air,
every breath adds to the degree of impurity of the
air, so that the latter becomes more and more unfit
to breathe.   It becomes unfit to sustain life, not
only because it is robbed of the life-supporting oxygen, but because one of the gases which are expelled
from the body, by the breath, acts like a poison
when it is inhaled (re-breathed) again.

Ventilation. Ventilation means the furnishing of the
needed supply of pure air.   Nothing is
more important in the line of hygienic or health
precepts than this, that we should avoid the breathing of air which has been made impure by the breath.
To avoid this, it is necessary to have, at all times, a
proper interchange between the air of the room and
the pure, free air without.   No person can remain
long in a closed room without being injured by his
own breath.   There must be a place of escape for

the impure air and a place of entrance for the pure air. Since the impure products of the breath, being warmer and consequently lighter on being exhaled, rise toward the top of a room, they escape best from the opening of a window at the top, while pure air enters best through an opening lower down. But without attempting to describe any of the numerous plans of ventilation, let this precept suffice: Get from the abundance of pure air which God has provided, as much as you can at every breath, and avoid, as a poison, the inhalation of impure air.

THE VOICE.

The organs of the voice are so closely connected with the respiratory system that we will give a brief description of them here.

The chief organ of the voice is the *larynx*.

The Larynx. This is really an expansion of the upper end of the wind-pipe, as already seen. (*Refer to the manikin of the larynx above last lung section*). The structure of this wonderful musical instrument is very peculiar. It comes close up to the mouth, being directly connected with the bone at the base of the tongue. This (hyoid) bone with its peculiar "horns" is shown at (1) and (2) on the section. The frame of the larynx is composed of a number of cartilages. The largest of these is the *thyroid* cartilage (4), whose prominent projection in front is commonly called, "Adam's apple" (5). At (3) is shown a ligament, and at (6) a muscle which

unite the tongue bone and the thyroid cartilage.
Below, at (9) is a ring-like cartilage called the
cricoid (9), united to the thyroid by muscles (8).
These, with several other cartilages, form a sort of
box, along whose inner sides are stretched two pairs
of cords. The upper pair (13, *second section*) are
called *false* cords. The lower pair (12), are the
true *vocal cords*. These come more or less closely
together at the middle of the larynx, the slit or chink
between them (14) being called the *glottis*. Through
this glottis every in-going and out-going breath
ordinarily passes silently. But when the muscles
which regulate the vocal cords tighten up these
cords, while air is being expelled from the lungs, a
sound is produced, either high or low, according to
the degree of tension or tightness, to which the vocal
cords are drawn. The *epiglottis* (11) is a curious
lid-like cartilaginous covering at the top of the
larynx. It is attached at one end but loose at the
other. The object of the epiglottis is this: Since
the tube through which the food passes from the
mouth to the stomach lies immediately behind the
larynx and trachea, the act of swallowing would
evidently cause food particles to enter the larynx
and even the lungs, if there were no provision made
to prevent it. Such prevention is the work of the
epiglottis. The same muscular act which, in swallowing, causes the food to pass from the mouth into the
throat, also shuts the epiglottis down, like a trap-
door over the opening of the larynx, so that it passes

safely on into the proper tube which carries it to the stomach.

An Experiment. If you have learned to sing up and down the eight tones of the musical scale, you may easily perform an experiment on the action of the muscles which control the tension of the vocal cords. Sound slowly the syllables up and down the scale. You will feel a change in the contraction of the larynx muscles at every change of tone. Going *up* the scale, that is, to a higher and higher pitch, you will feel a tightening action of these muscles; coming *down* the scale to a lower and lower pitch, an opposite effect will be felt.

Speech. The action of the breath on the vocal cords which has just been described, produces vocal sounds of different *pitch*. Their *loudness* depends upon the degree or force with which the breath is forced through the glottis, just as a boy's more forcible blowing through his mouth-organ, or a cornetist's more intense blast of air through his cornet, causes a louder sound to escape from these instruments. But the sounds of higher or lower pitch, and greater or less loudness, as they come from the larynx are not speech. To produce the articulate sounds of language, the sounds which are made by the vocal cords are very much modified and variously shaped by the changing positions of tongue and mouth, which are produced in speaking.

This is illustrated in every word we speak, and you will find it to be an interesting experiment to utter slowly the sounds which compose some word, while noticing the changes which you make in the position of your mouth and tongue.

OUTLINE.

THE RESPIRATORY SYSTEM.

What?
- Lungs, very light and spongy, chiefly composed of air-cells.
- Trachea—commonly called wind-pipe.
- Larynx—upper part of trachea, and organ of the voice.
- Air-passages through mouth and nose.
- Chief respiratory muscles—the diaphragm and inter-costal muscles.

Where?
- Lungs—in middle of the chest.
- Trachea—between throat and lungs.
- Diaphragm—between chest and abdomen.
- Inter-costal muscles—between the ribs.

Why?
- To furnish oxygen to the blood.
- To expel impurities from the venous blood.

QUESTIONS.

What is the use of the respiratory system?
What are the chief organs of respiration?
Where are the lungs located?
Describe their structure.
What is the trachea?
What vessel brings the blood from the heart to the lungs?

Tell what takes place in the lungs.

Is the act of breathing performed by voluntary or involuntary muscles?

Why is this a wise provision?

What is meant by *inspiration?*

Describe the process.

What is meant by expiration?

Describe the process.

What is the lining membrane of the chest called?

What is the benefit of full and deep breathing?

What are the consequences of cramping the chest?

What is the effect of exercise on the lungs?

How is the air made impure by breathing?

Tell what you can of the necessity of ventilation.

Where are the vocal cords?

How are the sounds of the voice produced?

How is the pitch of the voice produced?

How are the lower tones produced?

What other organs assist in forming the sounds of speech?

What is the purpose of the epiglottis?

THE DIGESTIVE SYSTEM.

Where does the building and repairing material, which is delivered to all parts of the body by the blood, come from ? How is it prepared and how does it find its way into the blood-vessels of the circulatory system ? These are some of the physiological queries which come to us now.

The matter of the body all comes from **Body-building from Blood.** our food. To provide our bodies with such material, of good quality and of proper quantity, ought to be the main object of our eating. Not all of the food matter which we eat is useful in the body. So the means of separating the useful from the useless is necessary. Then, again, the useful parts of our food must be very much changed before they can be used by the building cells of the body. So the means for its proper preparation must be furnished. The system whose organs prepare the needed elements of the food for the blood, and separate the useless from the useful portion, is called the *digestive system.*

The preparation of food material for **Aid of the Muscular System.** the blood requires many operations. So the digestive system has a greater number of special organs than any other system of the body. We have seen how the movements and work of the circulatory organs depend upon the strength and prompt action of the muscles of the heart; also, how the respiratory system depends upon the muscular system in the steady and proper action of the breathing muscles. So, here, as we study the processes of the digestive system, we shall find how its work depends upon the muscles which are assigned to the duty of producing the necessary movements of its organs. Even the muscles of the arm and hand perform the very first act in the process of feeding the body, in properly bringing the food to the mouth·

The whole process of preparing food material for the blood is called digestion. The first step, that is, bringing the food into the mouth, is called *prehension.*

Work of the Teeth. The second step in digestion is *mastication.* This is performed in the mouth by the *teeth.* The mouth and the teeth, as used in chewing or masticating the food, have been called the *mill of the body.* The grinding in this mill is done by the muscles which move the jaws so as to produce a cutting, crushing or grinding effect upon the food by the teeth.

Number of Teeth. There are thirty-two teeth in the full set of a grown person. These are set in sockets of the upper and lower jawbones—sixteen in each jaw. Eight front teeth—four in each jaw—are called *incisors*, or cutting teeth. On each side of these incisors—above and below—is a *canine* tooth. The two upper canines are often called *eye-teeth*, and the lower canines, *stomach-teeth.* These twelve teeth—eight are incisors and four canines—separate or bite off a proper portion from the food which is brought to the mouth. Next to the canines are two *bicuspids*, on both sides of each jaw. Bicuspid means having two cusps. Then follow, as back teeth, three *molars* or grinders on both sides of each jaw. These twenty teeth—eight bicuspids and twelve molars—do the work of crushing or grinding the food to a proper degree of fineness.

Structure of a Tooth. As we have already learned, a tooth is not a bone. It does not belong to the skeleton. The teeth are instruments or organs of the digestive system. The structure of a tooth is an interesting study. To help us understand it, we are provided, in the Anatomical Aid, with the means of completely dissecting—that is, separating into its parts, the structure of a tooth. In the manikin of a tooth, the little projecting ridges at the top (1) are called *tubercles*. The portion above the gum (2) is called the *crown*. This is covered by a thin layer of *enamel*, the hardest material in the body. The tooth is mainly composed of a substance called *dentine*, or ivory (9). At (6) and (7) the roots of the tooth are shown. At (12) blood-vessels and nerves are seen to enter into the tooth. When an opening occurs in the body of the tooth, from decay, this nerve is exposed to the air and the action of food particles, and toothache is the result.

Work of the Salivary Glands. The third step in digestion is *insalivation*. Really, mastication and insalivation are performed at the same time. While the food is being chewed by the teeth, a liquid is mixed with it in the mouth. This liquid is specially prepared for this purpose, from the blood, by a number of organs called *salivary glands*. By the way, let us not forget that all the substances of the body are prepared from the blood. A *gland* is an organ which *secretes*—which means *separates*—some special or peculiar substance from the blood. So, one of the glands of the eye secretes *tears*, and the salivary glands secrete *saliva*. Three pairs of these glands are quite prominent. The largest pair is just below and in front of the ears. The second pair, in size, lies under the jaw-bone, and the third pair is under the tongue. These glands, between meals, furnish enough saliva to keep the mouth moist. But when food is taken into the mouth and chewed, or even at sight of something tempting to the taste, they furnish it in great abundance. It not only moistens the food so that it may be easily swallowed, but it also begins the process of changing the food material.

Work of the Pharynx and Œsophagus. The fourth step in digestion is swallowing. This is a much more familiar word than *deglutition*, which is the scientific name for the same act. When the food has been properly prepared in the "mill of the mouth," it is swallowed, or sent to the stomach. The cavity back of the mouth is called the *pharynx*. Between the pharynx and the stomach (*refer to body manikin*) is this tube (35) called the *œsophagus*. By the action of the muscles of the pharynx and the œsophagus, the food is moved into the stomach. This is *deglutition*.

Work of the Stomach. We have now traced the course of the food into the main organ of digestion— the *stomach*. Here its greatest change is to be produced. The position of the stomach should be well understood. Observe it carefully in this manikin (31). The stomach has three coats or walls. The outer coat is thin and smooth, and fitted for the protection of this organ in its contact with other organs. The inner coat—called the mucous wall—contains many little glands or cells, which secrete a substance which is very important in the process of digestion. It is called the *gastric juice.*

Use of the Gastric Juice. When food comes into the stomach the gastric juice is mixed with it. This intermixing is made quite thorough by the action of the muscles which compose the middle coat of the stomach. As long as there is food within it, these muscles keep up a churning motion of the organ. The result of the action of the gastric juice is that the food is very much changed in its nature and appearance. It is now called *chyme*, and the change which has been produced in the stomach, *chymification.*

Work of the Pylorus. At the right end or discharging opening of the stomach is placed a muscular valve called the *pylorus*. This name means *gate-keeper.* The pylorus is a door-keeper of the stomach. Such portions of the contents of the stomach which have been properly changed into chyme, it allows to pass out, but refuses passage to other portions. Much depends upon the faithfulness of this pyloric muscle. When from any cause it loses its power, or refuses to act, the food escapes from the stomach before it is prepared to enter the intestines, which is a form of indigestion which soon destroys life.

Work of the Liver. The *Liver*, as the manikin shows, is a very large organ overlying the stomach. It weighs from three to four pounds. It is both a blood-purifying and a secretory organ. As a secreting organ, it performs its part in the process of digestion by furnishing a substance called *bile*, which it sends through a duct or tube into the *duodenum*, or upper part of the small intestines, where it aids in further change of the chyme which has just passed into the intestines from the stomach.

Work of the Pancreas. Back of the stomach lies the *pancreas* (59). This organ furnishes a fluid called the *pancreatic juice*, which is also brought into the duodenum. The action of the bile, the pancreatic fluid and the intestinal juice, is to change the chyme into *chyle*, and to separate the useless or waste portion of the food. This waste portion is carried out of the body by way of the intestines, and the useful portion, having undergone all the processes of digestion, is now ready to be given to the circulatory system for transportation to every point of demand. How the chyle is transferred from the digestive organs into the blood will be shown in the next chapter.

Health of the Digestive System. Perhaps no system of the body is more carelessly or more frequently abused than the digestive system. No system of the body brings back upon the abusing offender a severer penalty of discomfort. Proper digestion is the very first condition of good health. Hence the hygienic principles referring to this system should be carefully learned and regarded.

Eating Too Fast. The injury resulting from eating too fast comes chiefly from this, that the processes of mastication and insalivation cannot be properly performed. Unless the food is properly chewed, and thoroughly mixed with saliva, its digestion in the stomach will be either much retarded or left incomplete.

Eating Too Much. The capacity of the stomach is limited. If it is overloaded, it can not thoroughly digest its contents. Besides, the gastric juice is also limited in quantity, and will not completely change into chyme more than a proper portion.

Eating Too Frequently. Between the digestive operations of the stomach, it needs intervals of rest. If food is taken too frequently, it loses its vigor, and soon fails to perform its work in a healthy manner.

Eating Indigestible Food. Some articles of food, though very tempting to the taste, are very "trying" to the stomach. It is plain, that, if these are too frequently eaten, the stomach's action will be greatly impaired. If eaten at all, they should be sparingly mixed with more digestible food.

Exercise. Gentle exercise is very helpful to digestion. But violent exercise, either just before or after a meal, is quite as injurious. A cheerful state of mind is very helpful in keeping up a healthy action of all the digestive organs.

OUTLINE.

THE DIGESTIVE ORGANS.

WHAT?
- Teeth—32 in full set.
- Salivary Glands—three pairs.
- Muscles of Pharynx and Œsophagus.
- Stomach.
- Liver.
- Pancreas.
- Intestines.

WHERE?
- Teeth—in sockets of jaw-bones.
- Salivary Glands—located about the mouth.
- Pharynx and œsophagus — funnel and tube — between mouth and stomach.
- Stomach—under diaphragm in abdomen.
- Liver—overlying the stomach.
- Pancreas—lying back of the stomach.
- Intestines—filling lower abdomen.

WHY?
- Teeth — to masticate the food.
- Salivary Glands—to furnish saliva.
- Muscles of pharynx and œsophagus move the food from mouth to stomach.
- Stomach—to change food to chyme.
- Liver—to furnish bile.
- Pancreas—to furnish pancreatic juice.
- Intestines—to complete the work of digestion and separate the chyle from the waste matter.

QUESTIONS.

From what does all the body material come?

What processes are necessary to fit food for body nourishment?

What is the name of the system which performs these processes?

What is the first act in the process?

What steps take place in the mouth?

How many teeth in a full set?

Name the different kinds of teeth.

How is the food conveyed from the mouth to the stomach?

Describe the work of the stomach.

By what means is the gastric juice well mixed with the food?

What is the food, as it leaves the stomach, called?

Describe the action of the pylorus.

What does the liver furnish for the work of digestion?

What is furnished by the pancreas?

What is the chyle?

What are the consequences of eating too fast?

What results from over-eating?

What from eating too often?

What from eating indigestible food?

What are the effects of exercise on the digestion?

THE ABSORPTIVE SYSTEM.

In previous lessons we have learned how food is prepared by the organs of the *digestive system* and reduced to a fluid state. Let us also recall what we have learned of the *circulatory system;* how the blood is constantly making the circuit of the body, carrying food to every tissue.

The sixth chart of the Anatomical Aid shows most admirably all the organs and parts concerned in the two processes just referred to, and a careful study of this plate will enable us to understand their relation to each other. How, then, does the digested food get into the circulation? What provision is made for transferring it from the digestive system or *alimentary canal* into the blood-vessels? To answer these questions, we must first study how food is *absorbed*.

Absorption. The food, in its liquid form, must in the first place be removed from the stomach and intestinal tube. This is accomplished by a process called *absorption*. This work may be more easily comprehended by first referring to a similar process constantly going on in vegetable growths. When set in good soil and supplied with water, a plant will send out its small rootlets, whose little mouths will drink in (*absorb*) mineral substances from the soil, dissolved by the water. This liquid plant-food is carried to the sap (*vegetable-blood*) up through the trunk or stem, to nourish the parts of the plant. There is a similar provision for taking up the liquid food from the alimentary canal into the circulatory channels. There are little rootlets provided for this work whose action resembles that of the root-fibers of the plant.

Stomach-Wall Absorption. Really, the process of digestion begins as soon as the saliva comes in contact with the food in the mouth. That part of the food which is thus prepared for the circulation by the saliva, together with such liquid portions of the food as are ready to be taken into the blood at once, *do not pass out through the pyloric valve into the intestine,* but are at once absorbed through the walls of the stomach by the rootlets of the *gastric vein.* Refer to the sixth chart and observe how these rootlets of stomach veins are distributed all through the walls of that organ. These absorbents, after taking up or absorbing the prepared food portions from the stomach, collect in the *gastric vein* (39) which empties into the portal vein (60) which throws venous blood mixed with the absorbed food portions into the liver, where, after being acted on more or less by the liver, it leaves that organ through the *hepatic vein* (52) to be emptied into the great ascending vena cava (51), which carries it to the right auricle of the heart.

Intestinal Absorption. The food which passes out of the stomach through the pyloric valve to be farther digested in the intestines, is taken into the circulation in two ways. Food solutions which are not fatty, go by way of the liver, while fatty substances go by altogether a different transfer route. Let us first follow that portion which goes from the intestines to the heart by way of the liver.

Referring again to the chart last mentioned, notice how the rootlets of the *mesenteric veins* (58) are distributed through the walls of the intestines. As the gastric veinlets collect food solutions from the stomach, so these mesenteric veinlets collect digested food substances from the intestines, to be carried by the *portal vein* (60) into the liver and from thence to the heart.

Intestinal Villi. For the double absorption which goes on in the intestines, there is a special provision made in the structure of these organs. The inner wall or coating of the small intestines has a velvety or plush-like appearance. This is due to the myriads of little hair-like projections which hang down from the inner walls and point toward the center of the tube. These small cones or fingers are called *villi*, a word which signifies *hair-like bodies*. These villi are very numerous, covering the intestinal lining as with a coat of hair. Observe that in the middle of the chart a portion of the intestine is represented as cut open to show these villi points within. These villi, which dip into the liquid contents of the alimentary canal, are not themselves the absorbents, but they contain the absorbing rootlets which take up the food and start it on the way to the heart.

Structure of the Villi. The lower right-hand figure on the sixth chart gives us a beautiful idea of the structure of the villi, as seen through the eye of the microscope. A dozen or more villi are shown in the figure. Three of them are shown as cut open. The walls of each villus is curiously laid up of many cells (epithelium). Within are blood-vessels, arteries and veins. These are the mesenteric absorbents which have already been mentioned as collecting into the portal vein.

Lacteals in the Villi. The microscopic view also shows a single tube, with rootlets, extending lengthwise through the center of each little villus, surrounded by the meshes of the small veins just described. This tube is a *lacteal*. *Lac* means milk, and it is owing to the milk-like appearance of their contents that the term *lacteal* is applied to these absorbing vessels.

Use of the Lacteals. The lacteals absorb the fatty portions of the digested food. After they emerge from the villi, they are called *chyliferous vessels*. These pass through numerous *lymphatic glands* (mesenteria) (d), and finally empty into the *chyle receptacle* shown on the main figure at (24).

Thoracic Duct. The chyle receptacle is a sac-like expansion of the lower end of the *thoracic duct* (25), which is about as large as a slate pencil or a goose-quill. This carries the chyle upward in front of the spinal column and behind the oesophagus (I). At its upper end it bends forward and downward, something like the crook of a walking-cane, and pours its contents into the left *subclavian vein* (50), from which point it soon reaches the *great descending aorta* (27) which empties it into the right auricle of the heart.

What Has Been Done. Thus we have traced the different substances of which the digested food is composed, by different routes to the same cavity of the heart, there to mingle with each other, and with the impure blood collected from all parts of the body. We have found the *viaducts* which span the gap between the digestive and the circulatory systems.

Assimilation. The food is now in charge of the great carrier, the blood, which is propelled by the heart to the lungs to be purified, then returned to the left side of the heart, whence it is driven out through the great aorta and its numerous subdivisions to all parts of the system. Now, just as a freight agent at a railway station takes from a stopping train just such freight as is intended for his particular station, so every tissue of the body selects just such parts from the blood as it can appropriate to its use or utilize in repairing itself. The bones take from the blood such material as will make bone-cells; the muscles will select such as will make muscular tissue. In this way every part of the body is nourished, and the wonderfully mysterious process by which each cell and tissue selects materials brought to it for its growth and development is called *assimilation.* This term is appropriate, because it means *making like.*

Recapitulation. So that the important processes which we have just been studying may be impressed upon our minds and remembered, let us recapitulate by tracing a morsel of bread and butter from the mouth to the tissues of the body which are nourished by it. "Bread and butter" may be considered as an almost perfect diet, since it contains nearly every ingredient necessary to sustain life. The bread represents one class of food, and the butter another class.

The Bread. 1. It is masticated in the mouth, and thoroughly mixed with the saliva, which begins the process of digestion.

2. By the action of the muscles of the pharynx and oesophagus it is forced into the stomach.

3. In the stomach it is acted upon by the gastric juice and most of it is digested.

4. It is forced through the pylorus into the small intestines.

5. Here it is absorbed principally by the *blood-vessels* of the villi.

6. The small veins now carry it into the portal vein, which empties it into the liver from below.

7. It passes out of the liver and is carried by the large ascending vein to the heart.

8. It is carried with the blood to the lungs, and is returned as red blood to the heart.

9. It is pumped by the heart to all parts of the body.

10. It is assimilated by the various tissues.

The Butter. 1. It is masticated with the bread.
2. It is swallowed.

3. It passes from the stomach to the intestines unchanged.

4. By the action of the bile and pancreatic juice, it is converted into *chyle.*

5. Being fatty substance, it is absorbed chiefly by the lacteals.

6. The chyliferous vessels carry it through the lymphatic glands into the chyle receptacle of the thoracic duct.

7. The thoracic duct empties it into the sub-clavian vein from whence it soon reaches the right auricle of the heart.

8. From this point, the two kinds of food travel together in the current of the impure blood, to the lungs; then with the purified blood back to the heart; thence to the out-posts of the body, carrying nourishment wherever needed.

Lymphatic Absorption. Besides the absorption processes already described, by which the nutritive food substances are transferred from the digestive system into the channels of the blood, there is another very important absorptive process carried on in all parts of the body. The vessels by which this is done are called the *lymphatics*.

Structure and Distribution of the Lymphatics. The lymphatics commence as very small absorbing rootlets or tubes in all parts of the skin, in the lining membranes of passages and cavities within the body, in the covering membranes of all the internal organs, and especially at all points where secretions are apt to accumulate. In gathering from their numerous points of beginning towards the more central parts of the body, the lymphatics pass through many little glands or knot-like bodies, which vary in size from that of a pin-head to an inch in diameter. These are called *lymphatic glands*. Some of the lymphatics, from the upper part of the body, empty directly into the great veins near the heart. Most of them, however, are gathered into larger vessels through which they finally empty their contents into the thoracic duct.

Lymph. The fluid which the lymphatics absorb is called *lymph*, which signifies *transparent fluid*. Lymph is found widely distributed throughout the whole system. In composition, it strongly resembles the plasma of the blood, and contains minute bodies or corpuscles resembling the white corpuscles of the blood; these are called *lymph globules* or *lymph corpuscles*.

Origin of Lymph. This fluid is supposed to be mostly worn-out material gathered from all parts of the body. It probably consists of portions of blood ingredients which have oozed through the walls of the arteries, veins and blood capillaries, together with certain products of the exchange of old for new material, which is continually going on in the body. These substances are gathered up by tiny vessels, and, after being worked over, in a manner not well understood, they are capable of further use in the body. Thus we see a wise economy in allowing nothing to go to waste which can in any way be put to further use. This reminds us of the economy practiced in sifting coal-ashes taken from our stoves and furnaces, saving therefrom such partly burned coal as may be capable of giving off more heat if put upon the fire again.

The Lymphatics a System of Drainage.

The functions or work of the lymphatics of the body may be regarded as similar to those of tiles, or drain-pipes, which farmers so frequently lay in wet, swampy lands for the purpose of carrying off the surplus water. The water soaks into these tiles, which carry it off under ground, thus drying the field. Likewise, the surplus fluids which collect in all parts of the body are absorbed by the lymphatics, the drain-pipes of the body, which unite, forming larger vessels, which empty into the great veins, and especially into the thoracic duct, with the contents of which the lymph reaches the heart.

Use of the Lymphatic Glands.

It is not definitely known what the work of the lymphatic glands is. But it is probable that they renovate, or *work over* the "second-hand" and surplus material brought to them by the lymphatics, and that the *lymph globules* originate in them. Whether this is true or not, there can be no doubt that these glands are essential to health; because, when they become hardened or inflamed, as is often the case in persons of a scrofulous tendency, health fails, and the patient grows thin and emaciated, even though his diet may be of the proper kind and quantity.

The Lacteals a Part of the Lymphatics.

The *lacteals*, which we have considered in connection with the absorption of the food from the alimentary canal, are a part of the lymphatic system. They constitute that portion which begins in the villi of the intestines. When the process of digestion is completed, they serve as *drain-pipes*, like the lymphatics, in the system at large. Their special work, however, is that in connection with the absorption of the fatty food through the walls of the intestines.

The Lymphatics Compared with the Blood-vessels. We have learned how the blood "circulates"; how it starts *from* the heart, and after making the complete circuit, is brought back *to* the heart again. We have also learned that the blood-vessels both give off tissue-making substances and take on in exchange, waste and worn-out material, which they carry away. In contrast with this, the lymph does *not* "circulate". It is carried *toward* the heart, to enter the life-giving stream—the blood. In the lymphatic system there are, therefore, no vessels to correspond with arteries. Again, the lymphatics collect worn-out tissues, but give nothing in return.

Other Functions of the Lymphatics. The work of the lymphatics is not confined to the absorption of food from the intestines and collecting surplus and waste material from the system in general. Certain other phenomena, all of which are of interest to us, are due to the absorbing power of these vessels. For instance, when a poisonous substance is placed upon the skin, the lymphatics at once *absorb* it and carry it into the circulation. "It is by the action of the lymphatics that medicines which are now frequently simply injected under the skin, reach the blood. Even food substances have been thrown into the circulation in this way." The lymphatics of the lungs take in the poison of disease and diffuse it throughout the system. When the appetite fails, during long-continued illness, life is sustained by the unconscious consumption of one's own flesh, the fatty and more fluid matter being absorbed by the lymphatics and carried into the circulation. In a similar manner, as we shall learn farther on, the poisonous nicotine of tobacco is absorbed in the lungs, and the system poisoned. Thus we see that these vessels, which are ever active, take up, indiscriminately, foods, poisons, medicines, or the waste of worn-out material.

Importance of Healthy Lymphatic Action. We can scarcely realize how much of our health and comfort depends upon an active condition of the absorbent system. When we remember how many special glands there are in the body and what an enormous amount of different secretions they produce, and that when such secretions have served their purpose, or have accumulated in too great quantity they must be drained away by the lymphatics, we may understand, in a measure, that the failure of the lymphatics to do their work well is a serious matter. Thus dropsy of the brain and dropsy of the heart are but a few examples of the consequences resulting from a failure of the lymphatics to carry off secretions which, though they have a very important service to render, must not be allowed to remain or accumulate after their work is done. So we have an illustration of lymphatic service in the case of a common blister. Though a large amount of fluid may accumulate, so that the outer skin is very much raised by it, it may wholly disappear without coming to the surface through a break in the skin, because the lymphatics in such case, absorb the watery-like contents of the blister. Likewise, tumors, and many forms of skin diseases and eruptions are evidences of imperfect lymphatic action.

OUTLINE.

WHAT? {
Food is transferred from the digestive organs into the circulatory system by the process of absorption.

There are two distinct transfer routes or methods.

Surplus fluids and worn-out tissues are collected by absorption and carried into the circulation. Some of the "second-hand" material is renovated or "made over" and again sent with the blood through the body for use.
}

WHERE? {
Absorption takes place — In the walls of the stomach,
In the intestines,
In the tissues of the skin,
In the cells of the lungs,
At the places where gland secretions accumulate, and
Wherever blood-vessels are found.
}

WHY? { To collect and transfer food and other material.

QUESTIONS.

What is absorption?

How do plants illustrate the manner in which food is absorbed?

What is there in plants to correspond with the blood in animals?

Where does food absorption begin?

By what is this first absorption accomplished?

By what route does the food which is absorbed from the stomach reach the heart?

Where does food absorption next take place?

What gives the lining membrane of the intestines its smooth, velvet-like appearance?

How many kinds of absorbents are in the intestines?

Describe the structure of the villi.

What are the mesenteric veins?

What is the portal vein?

Trace the food absorbed by the mesenteric veins from the intestines to the heart.

Where is the hepatic vein?

Into what greater vein does it empty?

What are the lacteals?

Describe their work.

Of what general system are they a part?

Through what glands do the lacteals pass?

Into what do they empty?

Why are they called chyliferous vessels?

What is chyle?

Describe and locate the thoracic duct.

What is its lower enlarged portion called?

What is meant by assimilation?

Trace a mouthful of food from the mouth to the tissue in the body.

What is lymph?

What is its origin?

What vessels carry it?

To what may they be compared?

In what respects do the lymphatics differ from the blood-vessels of the general circulation?

Whither is the lymph carried?

What are the lymphatic glands?

Why should we be careful not to touch poison ivy?

What danger is there in breathing the air of a sick chamber?

When a squirrel or other animal hides away to sleep all winter, without eating, how is life sustained?

Why is medicine sometimes injected under the skin, and how is it rendered effective?

Name all the vessels which constitute parts of the absorptive system.

THE EXCRETORY SYSTEM.

We have learned in the preceding chapter that some of the material of the body, when it has served its purpose in the structure of a certain organ or tissue, may be made serviceable in some other part or tissue, and that such second-hand material is gathered up and returned to the blood by the lymphatics—which have been quite aptly called the "ragmen" of the body. There is, however, much waste matter which cannot be thus turned to further use in tissue-building. This must be regularly and thoroughly removed from the system in one way or another. But even some of this waste matter, in the very act of separating it from the body, is made to do service in some of the vital processes. Thus, the liver secretes *bile* from the impure blood which flows through it. This bile, while it is thrown into the channel of the intestines, renders, as we have seen, an indispensable service in the final digestion of the food. But a large part of the worn-out matter of the body is fit for no farther use whatever, and must be expelled; because, if allowed to remain in the blood, such matter would not only be useless, but an actual poison in the system.

This calls for a sewer system in the body, and such needed sewerage has been very well provided.

The Excretory Organs. The organs whose function it is to take from the blood such substances as cannot be utilized again for the same purpose, but which must either be changed into some other substance, or expelled from the system, are the *lungs, liver, kidneys* and *skin.* Each of these is suited to take from the blood a certain kind of impurities, and either elaborate them into some usable substance, or start them in their course leading from the body. The four organs above named, together with the *large intestine,* constitute the organs of the excretory system.

Excretion by the Large Intestine. In studying the digestion of food, we found the last step in the process to be the separation of the *chyle,* or useful food substance, from the waste matter, or useless food portions. As to the chyle, we have learned how it is transferred from the intestines into the blood-vessels. But the waste matter passes on (*see sixth chart*) from the first part of the small intestines (*duodenum,* 4)—where its separation from the chyle chiefly takes place—through the middle part of the small intestine (*jejunum*, 13), on through the *ileum*, (14), then past what is called the *ilio-caecal valve* (15), into the *ascending colon* (18), and through the *transverse colon* (19) and *descending colon* (20) of the large intestine out of the body.

Excretion by the Lungs. We have already learned of the excretory work of the lungs, in studying the purification of the blood in these organs. The exchange which takes place in the lungs at every breath is very important. For the purifying, life-giving oxygen which is taken into the blood at every *in*halation, the lungs unload from the blood a quantity of impurities at every *ex*halation. The substances which are thrown out from the blood, in the act of its purification in the lungs, are *carbon-dioxide, watery vapor*, and various other organic impurities. How are these substances formed in the body, and how do they get into the blood?

Carbon-Dioxide. The chief excretion from the lungs is *carbon-dioxide*, or, as it is more commonly called, *carbonic acid gas*. A large part of the tissue of the body is composed of the elementary chemical substances called *carbon* and *hydrogen*. Now the oxygen carried with the current of the blood through the system, unites with the carbon of the worn-out tissues and forms this carbon-dioxide. This change takes place in the capillaries; so that while the arteries brought to the capillaries pure red blood, whose corpuscles were loaded down, like so many canal boats, with oxygen, the veins carry away from the capillaries impure and darker blood, in which the corpuscles carry carbon-dioxide instead of free oxygen, as a cargo.

Nature of Carbon-Dioxide. The "carbon" and the "oxide" in the name of this gas have reference to the carbon and the oxygen which unite in its formation. The prefix "di" in the last part of the name, has reference to the fact that two atoms or smallest chemical particles of oxygen unite with one atom of carbon. So when the chemist writes the name of this gas by what he calls a sign or symbol, he writes CO_2. This symbol and name distinguishes the gas which we are learning about from another very poisonous gas in which only one atom of oxygen unites with an atom of carbon, so that the chemist writes its symbol CO, and calls it carbon-*mon*oxide. This gas you have no doubt often seen burning with a pretty blue flame over coal which has been freshly put upon the fire in the stove. Strange that while carbon-monoxide burns, carbon-dioxide will promptly put out a fire.

Carbon-dioxide will not only fail to support life, when inhaled, but it acts upon the body as a deadly poison. Its destructive nature may be illustrated in many ways. Since it is about half again as heavy as air, it is apt to settle in low places, such as pits and wells, in dangerous quantity. So it has frequently happéned that men who have gone down into wells without first testing the air within them, have been suddenly overcome and fallen lifeless to the bottom.   It is never advisable to descend into a well before having lowered into it, to the bottom, or water-level, a lighted candle.   If the candle continues to burn, it may be concluded that the well is sufficiently free from carbon-dioxide to be entered with safety.   But if the candle goes out, it is suicidal to descend into the well; for where a candle will go out from the presence of carbon-dioxide, or for the want of oxygen, a man's life will go out.

Watery Vapor. How is the watery vapor which escapes from the lungs produced in the body? It is formed by the union of oxygen with the *hydrogen* of the worn-out tissues.   The two substances unite in the proportion of two atoms of hydrogen to one of oxygen.   The chemist symbolizes this substance H_2O, and calls it water, or watery vapor. Such vapor of water escapes with every out-going breath.   Ordinarily it is not visible, but in cold weather it may frequently be seen escaping in the form of a mist from the nostrils or mouth, or collecting on the cold window-panes of a room containing a number of persons.   It has been carefully estimated that about one or one and one-fourth pounds of water is daily given off with the breath of a man.

Other Products of the Breath. Besides carbon-dioxide and watery vapor as the chief and constant products of the breath, there are always more or less of other impurities unloaded from the blood through the lungs.   This fact is often very disagreeably proved to us by the offensive odor of the breath of other persons. To show experimentally that other substances, such as particles of animal matter are contained in air once breathed, let the contents of the lungs after a full inspiration be breathed into a bottle and corked up. The impure matter excreted with the breath will decompose and soon give off an offensive odor. Since neither carbon-dioxide nor watery vapor have any odor, the experiment proves the presence of other substances, and it is these substances, in variable kind and quantity, that give the breath a more or less offensive taint.

Farther Test of the Breath. To prove experimentally the deadly nature of the products of the breath, breathe into a glass jar, after having held the breath in the lungs for some time. If a lighted taper or wax candle be lowered into the jar containing this exhaled breath it will go out, showing that there is not enough oxygen in the air of the jar to sustain the life of the candle flame. Just as surely would the life of a man go out if he were placed in a vessel sufficiently large, and containing the same kind of exhaled breath mixture.

Ventilation. We have before referred to the importance of thorough ventilation. What we have just learned of the poisonous character of the excretions of the lungs, gives us occasion again to refer to the subject. Air which contains four or five parts in a hundred of carbon-dioxide is very poisonous, and even a much smaller percentage of it is quite injurious. Besides, there are various other air-polluting causes. The dead, cast-off material of the body, which, as we shall soon learn, escapes in great quantity from the skin, will, if cleanliness is not strictly observed, and proper ventilation carefully practiced, accumulate in the air of the room, to be inhaled by the occupants. Is it, in view of these facts, any wonder that the pupils in many school-rooms complain of headache, dullness and disinclination to study, or that in crowded, poorly ventilated churches, the preacher seems dull and the sermon uninteresting, while the people of the congregation yawn, or even fall asleep under the depressing influence of the air which is so breath-poisoned that the very flames of the lamps are sickly and threaten to die?

Excretion by the Liver. The secretion of bile by the liver has already been mentioned. In this production of bile, the liver acts as a purifier of the blood. It is a well established fact, that in case of a diseased liver, when that organ fails properly to perform its work of secreting the bile, the substance of which thus remains in the blood, a disease known as *jaundice* appears, and if this disease is not checked, the person dies with symptoms of poisoning. So the liver is both a *secretory* and an *excretory* organ.

Excretion by the Kidneys. This brings us to the study of a pair of organs which have not before been described, namely, the *kidneys*. These are two bean-shaped bodies, a little more than half as large as the closed fist. Refer to the last section of the body manikin, where their location in the back part of the abdominal cavity,—one on each side of the spinal column—and also their size, shape, color and structure are very plainly shown. These dark-colored glands have a very important function to perform. They cannot delegate their work to any other organ of the body, as is the case with some of the other glands. They alone can perform the work assigned them. Hence, when diseased, their work is not done, and sickness follows.

Work of the Kidneys. The special work assigned to the kidneys is to separate from the blood which is brought to them, a substance called *urea*. This is a very poisonous matter, which, if not removed from the body by the healthy action of the kidneys, will accumulate, and finally cause death. The whole liquid excretion of the kidneys is called *urine.*

How the Kidneys work. The *renal arteries* (g) constantly carry to the kidneys a portion of the blood, which passes through the capillaries of the kidneys, as seen by turning back the first section of the *right* kidney of the manikin. The filtered blood is again collected by the veins, and carried by the *renal vein* (55), to the large veins leading to the heart.

The second section of the *left* kidney of the manikin, shows another interesting view of the peculiar structure of the inner part of this organ. Into this inner chamber or reservoir, as shown, numerous pyramid-like bodies project (c). Near the points of these little pyramids are little openings. As the blood comes through the capillaries into these pyramid bodies, the urine, carrying in solution the *urea*, is strained from it and trickles through the end-pores into the central reservoir. From there it escapes through the *ureters* (56), into the *bladder* (57), and from thence through the urinary channels out of the body.

The Skin. We now come to the study of the last of the great excretory organs, namely the *skin*. Before we are prepared to understand how the skin performs its sewer work, we must give some attention to its general structure. The skin serves a number of important purposes. In the first place, it forms a protective covering for the whole outer body. It is interesting to observe how the structure of the skin is adapted to the protection of the parts which are variously exposed. Thus, over parts exposed to pressure and friction, the skin is thick and tough; over parts liable to changes in size, it is quite elastic, and over parts which are specially delicate; it is covered with a mat of hair. Thus, we see that it is everywhere adapted to the purpose of protection.

In the second place, the skin is the general organ of touch. For this purpose it is abundantly supplied with sensitive nerves, which, in some places, are more abundant and delicate than in others.

We shall also soon be prepared to understand why it is that the skin is considered to be, to a great extent, the regulator of the heat of the body.

Like the liver and the kidneys, the skin is both a secreting and an excreting organ. The proper performance of these functions, on the part of the skin, is quite essential to health and life as we found it to be in the case of the liver and the kidneys.

Structure of the Skin. The skin is a much more complicated structure than one would first suppose. So closely and delicately are numerous various organs interwoven in the web of the body garment, that it requires the aid of the microscope to help us understand how it is constructed and what it contains. Such a microscopic illustration of the structure of the skin is furnished us on the sixth chart, upper left-hand corner.

First, overlying the true skin is the *cuticle* (a). Sometimes this is called the *scarf-skin* or *epidermis*. It seems to be intended for a special protective covering for the real or true skin. It is this cuticle or outer skin which is raised by a blister. No blood flows from it when it is pierced or cut, because it has no blood-vessels. Neither does it give us pain when it is torn or wounded, because it contains no nerves. The figure shows its real structure; for, as the microscope shows, it is made up of small, flat scales overlying each other in rows of layers like the shingles on a roof. These scales forming the scarf-skin are constantly being formed from the true skin below. As they are thus newly added to the under surface of the cuticle, they are constantly falling off from the outer surface. This shedding off of cuticle scales from the body, usually goes on quite imperceptibly. Sometimes, however, these cuticle scales accumulate and afterwards come off in masses. When this happens anywhere on the general body, it is called *scurf*. When it occurs on the head it is called *dandruff*. Another peculiarity of the cuticle is that over such parts of the body that are much worn, it grows very thick and tough. This is another of Nature's numerous provisions for the special protection of endangered parts of the body. Every boy knows how tender the soles of his feet are when he makes his first bare-foot venture out-doors in the spring. But Nature soon meets the new want of a thicker skin-soling, by accumulating a cuticle layer of more than double thickness and of much greater firmness, so that the feet soon become quite insensible even to the roughest walk.

Complexion Cells. As the figure at (b) shows, the lower layers of the cuticle have a decided color, while the outer layers are almost colorless. This deeper part of the cuticle, lying next above the true skin, contains the coloring matter which determines the complexion of the person. The negro is black because the *pigment cells* — as the lower layer of cells in the cuticle are called — contain very dark granules. In the cuticle of the Indian, these pigment cells contain red or copper-colored granules. So different persons of the same race — as the white race, for instance — have darker or lighter complexions, according to the color tint of the under cuticle cells. The strong rays of the sun have the effect of darkening or "tanning" the complexion, either by causing an increase in the number of these color granules or by deepening their tint. Sometimes the pigment or coloring gathers in spots. These spots are called *freckles.*

Blood-vessels. Close under the cuticle lies the true skin with its wonderful variety of parts. The arterial system supplies it with blood through such a closely woven web of capillaries that one can scarcely prick the true skin anywhere with a needle-point without drawing blood from a capillary. The figure shows (f) how the smaller blood-vessels branch out from the arteries and are then re-gathered to form the blood-returning veins.

Nerve-loops. In a similar way nerve lines come quite near to the surface of the true skin, and end in little papillæ or nerve loops specially fitted to be extremely sensitive to touch. Even through the insensible coat of the cuticle, the softest touch impresses these nerves sufficiently to cause them to carry a report of the impression to the brain.

Oil-glands. For the purpose of keeping the skin soft and tender to touch, a system of tiny oil factories is provided in its structure. These are called *sebaceous glands* (g). These glands are not equally distributed throughout the skin. They are quite abundant in the face. On the head they are so numerous that the oily substance which they secrete from the blood and throw out upon the surface is a sufficient natural pomade for the hair.

The Hair. Since the hair is really a part or modification of the skin, it is proper to mention it here. The figure shows how a hair grows from the skin (d). The part imbedded within the skin is called the *root*, while that without is known as the *shaft*. The farther details of the structure of a hair are shown in a separate figure under the figure of a section of the skin. The color of the hair depends on color granules in the hair-cells. The change of color to gray or white, in old age, is due to a decrease in the quantity of this hair-coloring matter.

The Nails. It is proper also in this connection to mention the *nails* of the fingers and toes as modified parts of the skin, provided for the special protection of the parts on which they grow. In structure, they are found to be cuticle cells very solidly packed together. The nail grows from the skin chiefly at its rear end and partly from below, so that as it grows forward it becomes harder and thicker, and is worn away or cut off when it protrudes too far beyond the toe or finger.

Fat Cells. Clusters of fat cells are also distributed throughout the tissue of the skin.

These give it a peculiar softness of structure and add to the grace of outline of the general surface of the body.

Skin Muscles. Numerous small muscles are distributed through the skin. Some of these are shown in the figure on the chart. They are sometimes called *hair-muscles*, probably because in some animals they seem to serve to make the hair, bristles or feathers, "stand on end." The horse makes good use of these muscles when he shakes his skin to drive off the flies. Some conditions, like cold air or intense fear, rouse these muscles to strong contraction, so that the skin is raised up in numerous little points called "goose pimples."

Excretion by the Skin. Having now learned about the general structure of the skin, we are prepared to inquire how it performs its important excretory work. The amount of waste matter which the skin carries out from the body is enormous. The matter which escapes from the skin pores is called *perspiration*. Observe, in the figure, numerous ducts or canals, opening out upon the surface. These are called the *perspiratory ducts*. They end in the skin in little coils. These are called *perspiratory glands* (h). Just as the bile is filtered from the blood by the liver, and the urine by the kidneys, so the perspiration or sweat is filtered or secreted from the blood by the perspiratory glands and carried to the surface by the perspiratory ducts. These sweat-glands and tubes are much more numerous in some parts than in others. The entire number on the whole body is supposed to be about two and one-half millions, and their total length, if they were placed in line, about three miles!

Ordinary Perspiration. Usually, the perspiration which escapes from the skin pores passes off as an invisible vapor; not, however, without leaving on the surface of the body more or less of a deposit of matter which does not evaporate with the liquid portion of the perspiration. This ordinary perspiration is always going on, even in the coldest weather. Since it is usually unobserved, it is often called *insensible perspiration.* It is always absolutely essential to health and comfort.

Sweat or Sensible Perspiration. When the amount of perspiration thrown off through the skin is much increased, as from vigorous exercise, or exposure to great heat, it is not evaporated from the skin surface as fast as it is produced, but collects in drops, or even little streams, on the surface of the body. This is called *sensible perspiration*, or more commonly, *sweat.*

Quantity of Skin Excretion. It has been estimated that on an average, about two pounds of waste matter is carried out through the perspiratory tubes each day. About ninety-nine parts out of one hundred, of this matter, is water. This, as already stated, passes off as vapor. The residue of one per cent. is left behind, on the surface of the skin, to escape therefrom as invisible dust, or to mingle with the material of the clothing, or to be washed away in the bathing of the body.

By the evaporation of the watery portion of the perspiration from the surface of the skin, the heat of the body is much reduced. This is on account of the well-known fact that when a liquid is changed into a vapor, it takes in much heat from surrounding objects. On this principle we sprinkle a room, on a hot summer day, to cool it. The sprinkled water passes into invisible vapor, and in so doing takes from the air in the room much of its heat, thus making it much more comfortable. We scarcely realize how much comfort we enjoy from the passing off of the perspiration of the body into vapor. When from any cause this process is arrested, a feverish heat and extreme discomfort is soon experienced. It is this condition which frequently leads people to say that if they could sweat more they would feel better. Thus the skin is an important regulator of the heat of the body.

Cleanliness. What we have just learned about the excretory work of the skin, certainly strongly urges the necessity of bodily cleanliness. It shows that frequent bathing of the body, as well as frequent change of clothing, are absolutely necessary to health; for we must not forget that what is once expelled from the body as an excretion, is afterwards a poison. This fact, in connection with what we have learned of the absorbing power of the skin, through its numerous lymphatics, is sufficient

explanation of the well-known truth that uncleanli-
ness of the body is usually followed by disease and
premature death.

OUTLINE.

WHAT?
> The organs of the excretory system are the
> *large intestine*, the *lungs*, the *liver*, the
> *kidneys* and the *skin*.

WHERE?
> The large intestine—in the lower abdomen.
> The lungs—in the chest.
> The liver—over, and to the right of the stomach.
> The kidneys—in the back part of the abdominal
> cavity, one on each side of the spinal column.
> The skin—covering the body.

WHY?
> The large intestine—to carry from the body the
> useless products of digestion.
> The lungs—to separate carbon-dioxide, watery
> vapor and other impurities from the venous
> blood.
> The liver—to secrete bile from impure abdom-
> inal blood.
> The kidneys—to filter urine from the blood.
> The skin—to collect impurities from the blood
> and carry them out through the *perspiratory
> ducts*.

QUESTIONS.

What different kinds of waste matter have we learned about?
What must promptly be done with such waste matter as is
entirely unfit for any farther use in the body?
What is meant by *excretion?*
What are the organs of the excretory system?
By which of these organs is the useless part of the food
carried out of the body?
What is the upper or first part of the small intestine called?
The middle part? The lower part?
What are the three divisions of the large intestine?

What substances do the lungs expel from the body?

Tell what you can about carbon-dioxide.

What proof is there that watery vapor escapes with the breath?

How can it be proved that other organic impurities escape from the lungs?

How can the deadly character of air which has been once breathed be shown?

What important hygienic precept do these facts suggest?

What causes the feeling of dullness and sleepiness in a school-room or church which is poorly ventilated?

What makes the lights grow dim in such a place?

What substance does the liver separate from the blood?

What is the consequence when the liver fails to do this work properly?

Why is the liver said to be both a secretory and an excretory organ?

Where are the kidneys located?

What excretory work do they perform?

What tubes connect the kidneys with the bladder?

Tell what you have learned of the different purposes of the skin.

What is the outer skin called?

Why can it be cut or torn without pain?

What does the microscope reveal about its structure?

What is meant by pigment or complexion cells?

What is meant by "scurf" and "dandruff"?

What causes the "tanning" of the skin?

What is the cause of "freckles"?

What proof have we that the skin is well supplied with blood-vessels?

What makes the skin so very sensitive to touch?

What provision is made for keeping the skin soft and smooth?

What are these oil-glands called?

What can you say of the structure of a hair?

On what does the color of the hair depend?

Tell what you can of the structure and growth of a finger nail.

Are muscles found in the skin?

What do you know about their use and action?

What is the excretion of the skin called?

What is the work of the perspiratory glands?

What is the work of the perspiratory ducts?

What is the difference between insensible and sensible perspiration?

How much matter passes daily from the body through the perspiratory tubes?

How does the perspiration escape from the surface of the skin?

Does it *all* evaporate?

What becomes of the residue?

What is the effect of the evaporation of the sweat on the temperature of the body?

Can you state the principle upon which this effect depends?

What do the facts learned about the excretion of the skin suggest in reference to cleanliness?

ALCOHOL AND THE BODY.

Value of a sound Body. The principal organs of the body, and the work or functions of each, are now quite familiar to us. Let us now consider briefly the necessity of guarding against anything and everything which would in any way impair the health of these organs, or interfere with them in the performance of their work.

If a grain of sand should find its way into the eye, inflammation would at once result; sight, the function of the eye would be interrupted. If, on account of some disease, the muscles of the heart should cease to contract and expand with their ordinary regularity, or stop their action entirely, the blood

·

would cease to circulate and life would end. Thus, the well-being of the body, yea, life itself, depends upon the healthy action of the various organs which are our servants in our body-house.

Does it not seem strange, then, that so **Abuse of the Body.** many thousands should still persist in abusing their bodies, which are made "in the image of their Creator?" Yet there are such who wilfully take into their system that which not only interferes with the healthy action of their bodily organs, but leads to certain death. More than this; they injure not only their bodies, but destroy their mental faculties, dethrone reason, bring misery and woe upon their families, and fail to accomplish life's ends. One of the most common wrongs against the body, producing such disastrous results, is the use of strong drink.

Under the name of strong drink are **What is Strong Drink?** included all liquids which contain *alcohol* in larger or smaller quantity. Such liquids will, when taken into the system, affect, more or less, all the organs and tissues of the body; and if the quantity or proportion of alcohol which they contain is sufficient, they will cause what is called *drunkenness* or *intoxication*.

Alcohol is a liquid so clear and color- **Nature of Alcohol.** less, that in appearance it cannot be distinguished from water. It has a strong odor, and a hot, biting taste. It is very inflammable. This may be easily observed by placing a small quantity in a shallow dish and applying a burning match, when it will quickly flash with a pale blue flame. Since it produces an intense heat, while it burns without smoke, it is very useful in the arts. It is much lighter than water, and boils, or passes into vapor, at a much lower degree of heat. Since its freezing point is much lower than any degree of cold which the atmosphere is anywhere liable to reach, it is used in thermometers where mercury would freeze. Alcohol is a powerful antiseptic, which means that it has such an effect on bodies which are subject to decay, as to prevent such a change. On this account it is much used for the preservation of animal bodies.

Another important property of alcohol is its water-absorbing power. It will take away water from any substance containing it. If the white of an egg, which is called albumen, be put in a cup, and alcohol be poured on it, the albumen will soon become white, hard and tough, as if cooked. In all these, and many other particulars, alcohol differs very much from water, which it so much resembles in appearance.

Sources of Alcohol. Alcohol can be made from many kinds of fruits and from most kinds of grain. But this is because such fruits and grains contain the substance from which alcohol directly comes. Really, alcohol is made from *sugar*, and the juices of nearly all fruits contain the sugar from which it may be produced. So, nearly all of the grains, as wheat, rye, barley, corn and rice, contain much *starch.* For example, the corn-starch which is sold by the grocer, is made from corn. But the starch of these grains is, under certain circumstances, easily changed into sugar. This is sufficient to explain why so many fruits and grains are sources of alcohol.

From Sugar to Alcohol. The change from sugar to alcohol is called a *chemical change,* because the chemical elements which make up a sugar molecule are separated so as to make *two substances,* each of which is altogether different from sugar. In other words, under favorable conditions, a molecule of sugar breaks up into *alcohol* and *car-bon-dioxide.* Alcohol is a liquid. Carbon-dioxide is a gas—the very gas we have learned about before, as being one of the products of the breath as it escapes from the lungs. The process of change from sugar to alcohol is called *fermentation.*

How is this Change Produced? To cause the breaking up of fruit sugar into alcohol and carbon-dioxide, as just explained, requires the presence and effect of what is called a *ferment.* Such a ferment is shown by the microscope to be composed of infinitely small germs or plantlets. Sometimes, as in the case of the ferment which is called *yeast,* these germs occur in *masses,* but besides such ferment masses, there are millions of ferment germs constantly floating in the air. The fact that fruit juices will undergo the change we have described apparently without the presence of any *ferment*, was long considered as quite mysterious, and led to the use of the name "*spontaneous fermentation*," which implied that such fermentation took place without the presence or influence of anything outside of the fruit juice which was changing. But the sharp eye of the microscope has discovered the real cause, namely, the invisible ferment germs which float in the air. These germs multiply rapidly in any substance in which they have started fermentation, and without their presence, or the presence of some mass ferment, fruit juices will remain unchanged an indefinite length of time.

How the Wines are Made. There are many kinds of wines, such as cherry wine, currant wine, gooseberry wine and rhubarb wine. Fermented cider, or, as it is commonly called, hard cider, is really apple wine. These wines are all made from the juices of the fruits after which they are named, by simple fermentation. Such fruits all contain sugar dissolved in their juices. When any of these juices, as currant juice, for example, is set in a warm place, freely exposed to the air, ferment germs soon start the change by which the sugar is converted into alcohol and carbon-dioxide. The carbon-dioxide being a gas, passes off in bubbles, and in so doing, makes the liquid froth. But the alcohol remains in the juice, which now tastes strong instead of sweet, and is called currant wine. By a similar process of fermentation all the wines are produced. When the process is to be hastened, a special mass ferment is added to the juice. All wines contain alcohol.

How the Beers are Made. The drinks of the beer family, including common beer, ale and porter, are made by a process called *brewing*. Brewing includes several steps which precede actual fermentation. Beer is chiefly made from barley, which, like other grains, contains starch. This starch of the barley grain is the starting point in beer making. It must be changed into sugar. This is done by causing the grain to "sprout." In sprouting, the starch of the seed is changed into sugar, which forms the first food of the growing germ. If you will take from the ground a pea or a pumpkin seed which is just "sprouting" and chew it, you will be convinced that there has been sugar-making going on in the seed. So the brewer exposes a large quantity of moistened barley to artificial heat, which soon causes the grain to sprout. When the change in the seed has progressed to the point of the greatest sweetness, which is well understood by the brewer, the grain is *roasted*, to kill the germ and stop the growth. If pale ale is to be made, the grain is but lightly roasted. For beer, it is quite browned, and for porter it is even charred. So far, this process is called *malting*.

Next the malted barley is crushed or "mashed" between heavy iron rollers. Then the mass is mixed in large vats with warm water. After some soaking,

the liquid portion is drained off from the mass.
This liquid is now called *"wort."*

Now comes the flavoring process. The "wort" is
mixed with hops or hop juice, and then thoroughly
boiled. This gives it a bitter beer flavor.

All these steps have been preparatory to the fermentation process. Yeast is now added as a ferment,
and the sugar change rapidly follows. The beer is
bottled or kegged while the fermentation is still
going on. By this means some of the carbon-dioxide which would otherwise escape, is held under
great pressure in the bottle or keg. This explains
the foaming or frothing of the beer when it is tapped
from the keg. All the beers contain alcohol.

How the Stronger Liquors are Produced. All the wines and beers are mild liquors
when compared with such drinks as
brandy, whiskey, rum and gin. This is
because the last-named liquors contain
a much larger proportion of alcohol. The process of
producing them is called *distillation.* This depends
upon the fact that alcohol passes from the liquid into the vapor state under much less heat than water
does; consequently the alcohol may be easily separated from any mixture which contains it, by applying
heat. As the alcohol passes off in vapor form it is
led from the heated vessel through a spiral tube.
This coil is surrounded with cold water, which makes
the alcohol vapor condense to the liquid form, and
flow from the tube into a receiving vessel. To get
perfectly pure alcohol requires repeated distillations,

and the use of certain substances which take from it the last traces of water.

"Brandy is distilled from the wines. Whiskey is distilled from fermented grains or potatoes. Rum is produced by distilling molasses. Gin differs from whiskey only in being flavored with juniper berries. The flavor qualities of these strong drinks come from the substances from which, or with which, they are distilled."

Proportion of Alcohol in Various Drinks. Brandy, rum, whisky and gin, all contain nearly fifty per cent. of alcohol. This explains why so small a quantity of these liquors will quickly make a man drunk. While the wines and beers contain a much smaller proportion of alcohol, their very alcoholic mildness makes them *dangerous.* Since their effect on the body, when first used, is not so severe as a drink of whiskey would be, their use is more likely to be repeated, and thus the habit of drink is stealthily but surely fastening itself upon the victim. From beer to wine and from wine to whiskey, the steps are easily and almost surely taken. Whatever may have been the liquor which was drunk, the effect is in every case the effect of the *alcohol* which it contains. Hence Dr. Richardson says: "In whatever form it enters, whether as spirit, wine or ale, matters little when its specific influence is kept steadily in view. To say this man only drinks ale, that man only drinks wine, while a third drinks spirits, is merely to say, when the apology is unclothed, that all drink the same danger." In proof that alcohol is a universal poison to animal life, the same writer says: "There is no animal that may not be affected by alcohol. At all events, I know of none. Some animals will swallow without injury substances that would be poison to man. A pigeon will take, without showing the slightest symptom, as much opium as would kill several men. A goat will swallow, without injury, a quantity of tobacco which would kill several men. A rabbit will swallow, without injury, a dose of belladonna that would kill several men. But neither the pigeon, nor the goat, nor the rabbit can swallow alcohol without being influenced by it much in the same manner as a man would be."

Adulteration of Liquors. It is very important to learn, here, that an unscrupulous desire for greater profits has led to the manufacture of so-called wines, beers and stronger liquors, which are only base imitations or adulterations. Drugs of various kinds are used to produce the imitated effects of color, sparkle and flavor. Thus the harmful effects of rank poisons are added to the injury done by the alcohol which such liquors contain.

EFFECTS OF ALCOHOL ON THE MUSCLES.

Not one of the great systems and organs of the body is proof against the effects of alcohol. Every organ and every kind of body tissue is more or less seriously injured by its use. The belief that alcohol gives vigor and strength to the muscles has long been entertained by many people. This is a great delusion which has numbered its victims by millions. The brief spell of excitement and apparent muscular vigor which follows the drinking of an alcoholic liquor, has been mistaken for strength derived from the alcohol. Really, it is the fitful reaction of the body in its effort to resist the effects of the intruding poison. It is a strong protest of nature against such intrusion. Instead of strength being imparted to the muscular system, as an effect of alcohol, it is now proved beyond doubt that strength is lost every time that alcohol enters the system. In every person the muscular system has a certain degree of elasticity, or vigor of contraction. A high degree of muscular elasticity or "spring" is *strength.* A low degree of the power of contraction is *weakness.* It has been shown by the most scientific experiments, performed by the most competent professional men, that the elasticity of the muscles is very much weakened by alcohol. One needs but observe the movements and work of a drinking man to be convinced of this truth. Such a man is unfit for any kind of work requiring strong and well-sustained muscular effort. Employers of laboring men understand this fact so well that they will refuse to employ persons who have been more or less muscularly "unstrung" by strong drink.

Alcohol has another directly destructive effect upon the muscles. It produces what is called *"fatty degeneration.*" Fat forms a very important part of the structures of the body, and is distributed in variable quantity throughout the different parts of the system. Besides forming a soft padding to the body, it really constitutes a reserve store of nourishment on which the body is fed when other food is wanting or cannot be eaten. All such fat, laid up in proper places and in proper quantity, is called *normal fat.* It is so called to distinguish it from fat which often accumulates as a consequence of disease. Muscle fiber, and various other tissues of the body, do, under certain conditions, lose their structure and change into fat-cells. This is called " fatty degeneration." This is very apt to set in as a result of the use of alcoholic drink—especially beer. It is most likely to occur in the muscles of the heart, kidneys and other prominent organs, as we shall see farther on.

The bloated appearance of many drinking persons has been mistaken by many as evidence of an abundance of muscle and vigor of body. None but the ignorant can longer be deluded with this idea; for any ordinarily intelligent observer must know that fat persons, as a rule, are weak in proportion to their fatness. This is always true when such fatness is an accumulation of the fat of degenerated tissue, brought on by the use of strong drink.

EFFECTS OF ALCOHOL ON THE NERVOUS SYSTEM.

We have learned how extremely delicate the structure of the organs of the nervous system is.

We have also learned that the derangement of the organs of this important system throws all the other organs and systems into confusion, and produces effects ranging from discomfort to disease and death. So after what we have learned of the nature of alcohol, we may well suspect that its effect on the delicate tissue of the nerves and nervous organs must be disastrous.

How does Alcohol affect the Brain? One of the first effects of the alcohol which is contained in the strong drink which has been swallowed, is to weaken the small blood-vessels which distribute the blood to all parts of the body. Ordinarily, these small blood-vessels have the power to resist a harmful increase in the flow of blood through them. But alcohol seems to partially paralyze them so that they lose the power to prevent an excessive blood-flow. Now, since the brain is very abundantly provided with blood-vessels, it is one of the first organs to suffer from alcohol. It becomes, as it were, flooded with blood. This often happens to such an extent as to produce *apoplexy* or *brain paralysis*.

Alcoholic Excitement. Though these more serious results do not always follow, alcohol always produces a sufficient rush of blood to the brain to throw it into excited, unnatural action. The brain, in turn, goads the organs which serve it into excited activity. So the eyes flash and roll wildly. The heart, sympathizing with the brain, throbs violently, thus doubling the first disturbing cause, by sending still more blood brainward. The mind is exceedingly active. All the movements of the body become more rapid and unsteady. The circulatory system, the respiratory system and the digestive system, are all partners with the nervous system in the wild spell of excitement under the whip and spur of alcohol.

Alcoholic Depression. After the excited condition just described, a state of exhaustion or depression always follows. This is because the effect of the alcohol on the blood-vessels gradually weakens. The wild beating of the heart gradually subsides and the flush of the face dies away. But, like a jaded horse, the heart now scarcely measures up to its usual work, so that the brain, instead of being crowded with blood, now scarcely gets a sufficient supply. So the mind settles into a condition of dullness, and all the organs of the body are drowsy and depressed.

Drunk. A person may pass through the physical excitement and the consequent depression resulting from taking strong drink, without really having been "drunk." Such a condition which just stops short of actual drunkenness is popularly spoken of as being "under the influence of liquor." When a man is drunk his condition is manifested by certain physiological derangements which are the direct effects of the alcohol contained in the drink he has swallowed. Not only has his brain been whipped up into a state of confusion, but the spinal cord has also become disturbed. So gradually the nerves which go out from the spinal cord lose their control over the muscles to which they go. First the nerve which goes to the lower lip and tongue fails properly to do its work. So the man stutters. Soon the nerves which control the muscles of the lower limbs fail to direct them. So the man shambles and staggers. By and by, the nerves fail to control the muscles of the eye. Squinting is the consequence. While this physical derangement exists, the mind is no less deranged, and the strongest passions of the man's nature, which, when sober, he is able to control, are now unchained and exhibited.

Dead Drunk. What do people mean when they say that a man is "dead drunk?" The condition which these words very aptly describe is the very next door to death. The victim is now *insensible.* Every organ has now yielded to the terrible strain of the drunken condition, and general prostration has set in. The man is apparently asleep. But it is not the sleep of health. It is a drunken stupor in which he is practically deaf, blind and insensible. Nearly the whole nervous system has suspended its operations, but fortunately *not quite all,* for the nerves which control the man's breathing and the beating of his heart, bravely and patiently hold out against the power of the tyrant that has invaded the body. So the man still breathes and his heart still beats; but that is all that separates him from death, until the flickering lamp of his life gradually recovers its burning.

Delirium Tremens. Sometimes the drunkard's course, from the cup to the grave, leads him by the way of a terrible condition which is properly called *drunken insanity* or *delirium tremens.* It is usually supposed that this condition is reached only after years of dissipation and drunkenness. Ordinarily, this is true. But reliable medical authorities tell us that persons of a particularly nervous disposition are sometimes attacked by this terrible malady when but small quantities of intoxicants are taken. Those who indulge in strong drink are never absolutely safe. It may attack them at any time.

The victim of delirium tremens is in terrible fear and anxiety. His mind is so completely disturbed, and his imagination so thoroughly aroused, as to cause him to think his best friends enemies, who would do him harm. He sees horrible sights, and hears noises which exceedingly alarm him. In his awful condition he raves and tears, cutting and biting himself like a madman. Not unfrequently, the victim dies under the spell, and thus escapes from his agony. Oh, that human beings should so abuse themselves as to bring themselves into such a condition!

Effects on the Brain's Structure. Medical authorities tell us that after the death of a hard drinker, more alcohol is found in the tissues of the various parts of the nervous system than in any other part of the body. It has been found in sufficient quantity in the brain to distill it from the tissue of that organ. Its abundance in the nervous tissue is probably due to the amount of water which the nerve tissue contains, and for which alcohol has a remarkable greed. Alcohol actually changes brain substance by its absorption of water. As we have learned before, if alcohol is poured into a cup containing the white of an egg, it will harden or coagulate it. The tissue of the brain is similarly affected and made less sensitive.

Alcoholic Softening of the Brain. Sometimes the effect of alcohol upon the brain manifests itself in another manner of degeneration of its tissue. When fat accumulates in the brain, from what we have learned to call "fatty degeneration," the result is known as *alcoholic softening.* This change of the brain's structure, under the influence of alcohol, is shown near the top of the ninth chart, where the left-hand figure represents the brain in health, while the middle figure shows the alcoholic effects on its internal structure. The right-hand figure shows the terribly congested (blood-flooded) and inflamed condition of the brain of a victim of delirium tremens.

Effects on the Nerves. The same effect which alcohol produces on the brain is noticed in the nerves. This effect is also illustrated on the chart (*near middle of left side*). The alcohol takes up the moisture in the nerve lines, leaving them more or less incapable of transmitting impressions.

There is on record an account of a man, who, in a drunken stupor, burned his foot almost to a crisp without becoming conscious enough to remove it from the camp-fire into which he had unconsciously placed it. His nerves were so thoroughly paralyzed by alcohol, as to fail to transmit impressions to the brain, even if that organ had been in a fit condition to receive the intelligence.

EFFECTS OF ALCOHOL ON THE CIRCULATION.

Alcohol, when swallowed, quickly finds its way into the circulation through its absorption by the roots of the gastric vein in the walls of the stomach. It soon reaches the smallest blood-vessels numerously distributed near the surface of the body. Its effect on the small blood-vessels, as we have seen, is to paralyze or relax their coats or walls. The result is that in this limp condition, the blood flows through them at a greatly increased rate. In consequence of this, the surface becomes warm and the flush of color comes to the skin from the presence of the excited scarlet flood. The surface heat thus produced led to the delusion that alcohol is a heat producer. But really it causes a loss of bodily warmth; for the heat which comes to the surface is radiated away and lost to the body, and there is no increase in the production of heat within to make good this loss. So, upon the whole, the temperature of the body is *reduced* by alcohol. This explains why drinking men generally complain of chilliness after the stage of alcoholic excitement has passed.

Effect on the Heart's Action. The blood-pumping action of the heart is due to the contraction and relaxation of its muscles. But it is more or less regulated by a certain degree of resistance which the small blood-vessels offer to the flow of the blood through them. In other words, the firmness of the walls of the small blood-vessels acts as a sort of a brake on the heart's action. But we have seen how alcohol weakens these blood-vessels so that they are relaxed and the blood rushes through them with unusual velocity and in unusual quantity. The consequence is that the heart beats more rapidly and with greater force. Thus the strain upon the heart is greatly increased, while the intervals of rest between the beats is diminished, which certainly is very injurious to that important organ.

This quickening of the heart's action after taking drink, led to the notion that alcohol is a *stimulant.* This is also a delusion. It has been thoroughly proved by the most competent experimenters that after such a temporary fit of alcoholic excitement, the heart is always more or less enfeebled. So really alcohol is a *narcotic;* that is, its stupefying effect much exceeds the temporary exhilaration which it at first produces.

Effect on the Heart's Structure. The heart is especially liable to "fatty degeneration," especially from the use of beer. Thereby the heart is not only weakened, but it is often completely fettered by a fatty coat gathering about it, which is known as fatty accumulation. The chart represents such a fat-encumbered heart. If the use of alcohol is con-tinued, the heart will finally succumb; its fibers will become relaxed; its cavities will become enlarged; it will entirely lose its power to contract, and death will result from paralysis or "failure" of the heart.

The muscular relaxation of the heart from alcohol is shown in the second heart figure on the chart (*from the left*). Sometimes the heart continues its efforts to expel the blood, even when the cavities have increased their capacity, and the walls have become thin and weak. Alcoholic rupture—that is, a breaking of the heart wall, as shown in the third and fourth heart figures—is then likely to occur. This, of course, means instant death.

Effect on the Structure of the Blood-Channels. As the use of strong drink is continued, the structure of the veins and arteries becomes more or less permanently changed (see figure, "Alcohol on Veins and Arteries"). The blood-shot eye and the redness of the drunkard's face are, as we have seen, caused by the crowded blood in the capillaries, under alco-holic excitement. When this surface inflammation occurs too often, it finally results in permanent capil-lary paralysis. This paralysis may occur quite generally over the body. Usually, however, it is more or less local, and seems specially apt to appear on the "toper's nose," as if nature were disposed to write the evidence of the alcoholic outrage on the body in as prominent a place as possible.

EFFECTS OF ALCOHOL ON THE RESPIRATION.

Effect on the Lungs. We naturally conclude that the delicate structure of the lungs renders them liable to harm from alcohol as it passes with the blood through these organs. This conclusion has been abundantly verified as correct. In consequence of the damage done to the structure of the air-cells by alcohol, the absorption of oxygen *for* the blood, and the separation of carbon-dioxide *from* the blood is not fully accomplished. Dr. Richardson says: "I found by experiment, that in presence of alcohol in the blood, the process of absorption of oxygen was directly checked, and that even so minute a quantity as one part of alcohol in five hundred of blood proved an obstacle to the perfect reception of oxygen by the blood."

Effect on the Blood Corpuscles. The consequence of such an imperfect oxidation of the blood is not only a loss of vigor to all the bodily organs, but even the red corpuscles of the blood are changed and distorted in size and shape, losing their roundness and becoming angular, on which account they meet with obstruction in the small capillaries. Thus the blood supply to any part is both deteriorated in quality and diminished in quantity.

Effects on the Breathing Muscles. Farthermore, the muscles which are chiefly concerned in breathing, like all other muscles, are weakened by alcohol. Sometimes they become affected by alcoholic fatty degeneration, especially in beer drinkers. To the extent of such effect on these breathing muscles, the blood-purifying process is hindered, and breathing becomes difficult. Many drinking men complain of "shortness of breath." This is due to an alcoholic crippling of their respiratory muscles.

EFFECTS OF ALCOHOL ON THE DIGESTIVE ORGANS.

Effects on the Stomach. The inner coat of the stomach is a very delicate mucous lining upon whose healthy condition the production of one of the most important digestive substances—the gastric juice—depends. It is a well-known fact that as soon as strong drink has been swallowed, the alcohol thereof begins its attack on this tender stomach lining by absorbing some of the water which is contained in its structure. This at once impairs the action of the gastric glands which lie in this coat, and whose work it is to secrete a proper portion of gastric juice for the complete performance of stomach digestion or chymification. A celebrated English physician has made the most conclusive experiments on this point. By exposing minced beef in bottles to the separate action of gastric juice and *water*, gastric juice and *alcohol*, and gastric juice and *ale*, he found, after a time, that while the beef which was mixed with gastric juice and water was properly affected and changed, that which was placed in the bottles containing gastric juice and alcohol or ale, was not at all being changed or digested.

Such interruptions of the digestive process and enfeeblement of the stomach's action, when frequently occurring, must seriously and permanently impair that organ, upon whose proper action so much of our physical comfort, and we may say, our personal happiness, depends.   The very common complaint of dyspepsia, among persons who use strong drink, is directly traceable to the effect of alcohol which has just been described.

The inner coat of the stomach is so abundantly supplied with capillary blood-vessels that the whole surface has a rosy tint.   This is nicely shown in the first figure of the seventh chart.   The liver naturally lies over the stomach, but in this figure it is represented as turned up, to show the appearance of a healthy stomach.

Inflammation. If alcohol were taken in an undiluted form it would seriously burn the mouth and throat.   The stomach would suffer in a similar way.   But in its most diluted form it has an irritating effect on all the lining membranes with which it comes in contact.   Inflammation tells the story of the unnatural condition of things. Such inflammation means unnatural heat, which is followed by unnatural thirst, which is the probable reason for the fact that the more liquor a man drinks the more he wants.

The second stomach figure shows the early stages of alcoholic inflammation.   It is true to life, being the picture of the stomach of a man who was a "moderate drinker."

Congestion. At a more advanced stage of the drinking habit, congestion, or permanent blood-crowding of the stomach sets in, as shown in the figure at the bottom of the chart. The blood vessels are very much expanded, and seem to have lost their power to contract sufficiently to resist the flood of blood brought to them.

Ulceration. By and by, if the terrible offense against the body is continued, ulceration follows. Ulcer sores form within, and even the walls of the stomach sometimes become riddled with ulcerous holes. Such a destruction of the stomach is represented in the figure on the eighth chart. The burned and chocolate-colored condition of a stomach in the last stages of delirium tremens is shown in the second figure, which speaks for itself.

The same inflaming and ulcerating effects are also produced by alcohol in the intestines, as the figures on the chart show.

Gradual, but sure Destruction. The faithful servant of the body, the stomach, tries, from the first, to adapt itself to the derangements caused by the use of alcohol as a beverage. Just as the cuticle of the hand thickens and hardens, when we use an ax or shovel, so the stomach, if it is constantly irritated by the presence of alcohol, becomes thick, tough and unnatural, and, consequently becomes better adapted for the purpose of a whiskey jug, but less for the purpose for which it was intended—to digest food. The blood-vessels of the stomach, having been stretched to their utmost, their diseased walls give way and ugly ulcers are formed, which cause great suffering. Unable longer to accommodate itself to the condition of things, the stomach gives up in despair. It can no longer retain, and much less, digest food. Pain and disease are now all that remain to the victim until death comes to his relief.

How small and seemingly insignificant was the beginning of this destructive process! A taste of wine, or a glass of beer. How certainly does the gratification of one thirst create the next! We would all do well to accept as our motto: "Touch not, taste not, handle not,"

Alcohol and the Liver. We have just seen how alarming is the effect of alcohol on the stomach. But the liver, the healthy action of which we have found so necessary in the food-digesting and blood-purifying processes, suffers fully as much from the use of alcoholic drinks.

After the alcohol has been absorbed by the veins, from the stomach, the first organ into which it is ushered is the liver. A proper secretion of the bile in the liver demands that the cellular structure of that organ remain unchanged. Alcohol causes a change of the liver cells to fatty tissue, and an enlargement of the organ follows. Its structure then becomes lumpy or knobbed, a condition which is known as "gin liver" or "hob-nailed liver." Compare its appearance under such conditions, as shown on the chart, with its looks when in health, as we have seen it on a previous chart.

Results. What is the result of a liver thus diseased? The answer is two-fold: First, the bile and liver sugar are not properly taken from the blood, and whatever poisonous matter is contained in the blood which should be removed with the bile from the system, must remain and will certainly prove destructive to health. In the second place, the digestion process is not furnished with the needed bile, and the work of preparing the food will be imperfectly performed.

Alcohol and the Kidneys. The effect of a continued use of alcohol upon the kidneys, is much like that upon the liver. In the kidneys the blood is constantly being filtered, and the poisonous urea is being taken from it and expelled from the body. If the blood carries alcohol into the cells of these organs, they will be irritated, inflamed, and sometimes destroyed. This is known as Bright's disease. Richardson says that seven out of every eight cases of kidney disease are traceable to the effect of alcohol. The kidneys are specially liable to fatty degeneration from beer drinking. Such a diseased fatty accumulation in and about these organs is represented on the chart.

TOBACCO

AND ITS EFFECTS UPON THE BODY.

The original home of the tobacco plant seems to have been our own America. Before the time of Columbus, it was not known to the people of Europe. A γ-shaped pipe which Columbus found in use by the natives of San Domingo was called "*tabaca.*" From this came the name, *tobacco.*

Tobacco is a strong narcotic, which means that its effects upon the body range, according to dose, from dullness of feeling and sleepiness, to stupor, convulsions and death.

Its Chief Poisonous Principle. Tobacco contains a substance called *nicotine,* so named after Jean Nicot, who sent seed of the plant to Catherine de Medici. This nicotine is a deadly poison. It may be extracted from the leaves of the tobacco plant as a dark-colored liquid, having a sharp, biting taste. It has been found that a few drops of this poison, when placed on the tongue of a dog, will cause death. It has also been found to form a considerable part of the crust in the bowl of an old tobacco pipe. Besides this poisonous nicotine, tobacco contains quite a number of other substances, many of which have an injurious effect upon the body.

The effects of nicotine upon the body are far-reaching and destructive. It weakens the action of the nerves, causing trembling; it deadens the nerves, producing paralysis. It produces functional disturbance of the heart, resulting in palpitation and irregularity of its action. It deranges the work of the stomach, causing dyspepsia and emaciation. It affects the eye, expanding the pupil and producing dimness and confusion of vision. In the ear it produces strange, ringing sounds, which in certain cases have been so annoying as to almost drive the victim to distraction.

The use of Tobacco Unnatural. That the use of tobacco is unnatural, is clearly shown by the nausea and the nervous and muscular weakness which it produces when its use is first attempted. The giddiness and headache which follows the use of the first cigar, or the first quid of tobacco, are Nature's means of protesting against the poisonous intrusion. Gradually, however, nature yields to the trespass, so far as her vigorous warnings are concerned; but the damaging effects of the use of the poisonous weed are not prevented by her reluctant surrender to the habit.

Cigarette Smoking. Injurious as tobacco smoking (in its common form) may be, smoking cigarettes is even more so. The poorest kind of tobacco is often used in making them, and poisonous substances are added to give them the proper strength and flavor. Opium, which is used in considerable quantities in this adulteration, is carried with the smoke to the lungs, to add its injurious effect to that of the poisonous nicotine. In view of the detrimental effect of cigarettes, it is not surprising that a number of States have enacted laws prohibiting the sale of cigarettes to young boys.

An Experiment. To show that the mucous lining of the mouth and air passages is subjected to a sort of tanning process, and thus has its ordinary functions impaired by cigarette smoking, a simple experiment may be performed. If we use a clean, white cloth or handkerchief, two or three double, and inhale the smoke of a lighted cigarette, and then force it from the mouth through the cloth, a brownish yellow spot or stain will be found on it, which consists of the poisonous nicotine and other ingredients contained in the smoke, and mixed with the moisture of the mouth. Deposits of this kind are made upon the walls of the air passages when cigarettes are smoked. That this must be injurious, no one can doubt.

Effects on the Respiratory Organs. The respiratory organs are directly and seriously affected by cigarette smoking. On the ninth plate of the Aid we have a series of views which will help us to understand the nature of the harm done, and by a careful study of these, we may, perhaps, all be led to resolve never to smoke cigarettes or cigars.

The trachea, or wind-pipe, in health, has the appearance shown at (1) and (2) in the larger figure. At (3) and (4) we are shown how this organ appears internally when it has become inflamed and irritated by the use of cigarettes. The inner structure of the lungs, with the subdivisions of the bronchial tubes and air-cells, in health, is nicely represented at (5). Here also we see the plump, full, well-formed lung, before it has become shriveled up by the effects of tobacco smoke and deposits. Contrast with this healthy condition the bronzed, hardened and contracted condition of the lung structure as shown at (6), where the air-cells are filled up, and the bronchial subdivisions are almost wholly obstructed by nicotine deposits.

The diseased condition of the heart chambers which frequently results from the smoking habit, is shown in the larger figure. The bluish figure, at the right-hand lower corner of the chart, represents a small portion of the lung more or less magnified. On this the deposit of nicotine sediment and other tobacco substances is clearly shown. Is it not plainly impossible that the blood-purifying process should be successfully carried on in lungs whose tissue is so much poisoned and obstructed?

Smoker's Cancer. Physicians are becoming convinced of the increased occurrence of a peculiar disease resulting from excessive smoking. This is called *smoker's cancer*. It usually occurs in the throat and often proves fatal. The appearance of the pharynx and the inner wall of the trachea, when affected by this disease, is shown in a figure on the chart.

It is now well known that the vigor of

Effect on the Circulation. the circulation is much diminished in persons who are much addicted to the use of tobacco. This results from the weakening effect of the nicotine poison on the motor nerves of the heart, by which this organ is made unable to pump the blood with sufficient force and in sufficient quantity to the more distant parts of the body. Such a decline in the vigor of the circulation of the blood predisposes persons to serious diseases, and so enfeebles them that they are unable to resist malarial influences or epidemics.

From a most excellent contribution to the *Kansas Medical Journal*, on "Tobacco—Its Effects upon the Eyesight," by Dr. F. B. Tiffany, we quote the following selected passages:

" Smoking or chewing in the young impairs growth, produces premature development and physical degeneration. Even the smallest amount of smoking is injurious to the immature. Smoking of cigarettes so universal among the youth, with its evil effects upon the organism is simply appalling. The evil effects of tobacco upon the youth should be pointed out by the medical profession; and, to my thinking, it would be well were there a law to be enforced by the superintendent and teachers of all public schools, prohibiting its use on the part of each and every pupil.

"Tobacco stifles the best mental impressions— ' blunts the keen edge of thought.' . . . Other things being equal, the best man wins in the life race, but the man with the dulled intellect, a tobacco heart, and a tobacco stomach, is certainly not on equal footing with him who is untroubled by the toxic (poisonous) influence of tobacco."

These are plain words from medical authority. In the article from which they are quoted, Dr. Tiffany shows the damaging effect of tobacco upon the eyesight, through many degrees of severity, from dimness of sight to total blindness or *tobacco amaurosis*. To show that his professional caution is not founded on mere theory, he cites nearly a score of remarkable cases of partial or total "tobacco blindness."

Experiments carefully made and described by Dr. De Caisne of France, show that one of the effects of nicotine and other alkaloids of tobacco is to produce a "rhythmical intermittency" of the beating of the heart. In simpler words, the regularity of the heart's beating is destroyed. This he found to be the almost invariable effect in boys from nine to fifteen. He also found that cigarette smoking has the effect to change the quality of the blood—robbing it to a large extent of its vital character. "This sluggishness of the blood invariably appears in the character or disposition of the boy in habits of laziness, stupidity, indisposition to apply the mind to study, predisposition to alcoholic stimulants, and sometimes complete moral and intellectual transformation, as well as physical degeneracy."

"A wealthy amateur had been selecting a micro

scope, using a drop of blood from his finger as a test, and was leaving the office of the optician with a cigar in his mouth. A professor of microscopy in a medical college happened just then to look through the instrument, which was still adjusted, and made a rapid mental combination, saying that the customer had but few months to live, unless he stopped smoking at once. But he did not stop. A sea voyage did not recruit wasted energy, and a few weeks later he died in Paris—his constitution breaking up, as the physician said."

To the foregoing conclusions drawn from the observations and experiments of competent professional men, the author will add that his own experience, as a teacher, is fully corroborative of what is represented in the preceding paragraphs as to the effect of the use of tobacco upon the body during the years of its growth. The smoking or chewing habit is never associated with perfection of physical and mental vigor in the same person. Tobacco stupor means weakened nerve power. Weakened nerve power means weakened thought power and weakened will power. Feeble thought power begets permanent mental imbecility, and feeble will power grows into moral depravity.

OPIUM.

Opium is a dangerous narcotic. Dangerous, because it beguiles its victim into a bondage from which release seems to be next to impossible.

This drug is extracted from the juice of the poppy. For this purpose the poppy plant is cultivated on a large scale in some eastern countries.

The active principle of opium is *morphine*. The name comes from *Morpheus*, the god of dreams and sleep.

The effect of opium upon the body is, first, to soothe the nerves, thus subduing the severity of pain, or even relieving it altogether.

In large quantities, it affects the brain peculiarly, fires the imagination and spurs up the organs of the body generally. But this temporary revel of the feelings is followed by a most depressed condition of mind and body.

Many pain-relieving and sleep-producing preparations owe their peculiar power to the opium which enters into their composition. Among these are laudanum, paregoric, soothing syrups, and Dover's powder. Not one of these should be used except under the direction of a careful physician.

"The victim of opium is bound to a drug from which he derives no benefit, but which slowly deprives him of health and happiness, finally to end in idiocy or premature death. Whatever the victim's condition or surroundings may be, the opium must be taken at certain times with inexorable regularity. The liquor or tobacco user can, for a time, go without the use of these agents, and no regular hours are necessary. During sickness, and more especially during the eruptive fevers, he does not desire tobacco or liquor. The opium eater has no such reprieves; his dose must be taken, and, in painful complications affecting the stomach, a large increase is demanded to sustain the system. If, in forming the habit, two doses are taken each day, the victim is obliged to maintain that number. It is the unceasing, everlasting slavery of regularity that humiliates opium-eaters by a sense of their own weakness."—HUBBARD.

ẶPPENDIX.

In the presentation of the subject of the circulation of the blood, for the use of ordinary pupils, it was considered best to give the essential facts, in as plain a form as possible, without encumbering the description with too many scientific terms, or with statements which would be of interest only to teachers and advanced pupils. For the latter class of readers of this book, we give the following additional facts relating to the circulatory system.

Blood. We have learned that the blood is the great nutritive fluid upon which our bodies subsist. We have also learned that in every round which it makes through the body it is made impure, and that on this account it must, once in every round, pass through the lungs for purification.

Blood is produced from solid and liquid food. There are about eighteen pounds of blood in the human body of average weight. The red corpuscles of the blood are the carriers of the oxygen which is taken in through the lungs. As this oxygen is exchanged in the body for carbon-dioxide, this gas is carried to the heart and thence to the lungs by the same corpuscles. The coloring matter of the red corpuscles is called *haemaglobine.* This, when separated from the blood, crystallizes. The peculiar shapes of these haemaglobine blood-crystals, as they appear under the microscope, are represented near the top of the sixth chart. It is an important fact that in the blood of different animals, these crystals have different shapes. Thus the microscopist is able to tell positively whether the blood which he is examining is the blood of a human being, cat, rat, squirrel or other animal.

The purpose of the white corpuscles, which are found in the blood in limited number, is not understood. The plasma, or fluid blood, carries in solution the nutritious elements of the food, which have been transferred to the circulatory channels from the digestive system.

The Heart Valves. The valve between the right auricle and the right ventricle is called the *tricuspid valve.* Between the chambers of the left side of the heart is the *bicuspid valve.* At the beginning of the pulmonary artery the *semi-lunar valve* guards against the backward flow of blood into the heart. The same effect is prevented by a valve at the beginning of the aorta.

The Arterial System. (*Refer to fourth chart.*) The pulmonary artery is the exception to the rule that the arteries carry pure blood. It conveys the impure blood from the heart to the lungs. The aorta is the great trunk of the arterial system. Very near its commencement it first gives

to the structure of the heart the *coronary* arteries
(33); then it gives off the *innominate* artery (18)
which, again, sends the *right subclavian* artery (20)
to the arm, and the *right carotid* artery (19) to neck
and head. Next, the *left carotid* (21) ascends from
the aorta through the left side of the neck to the
head.· The carotid arteries (19, 21) divide into
numerous branches, distributed to the jaws, (1,
external maxillary)—to the lips, (2, *lower labial*; 3,
upper labial)—to the nose, (4, *angular*; 5, *dorsalis
nasi*; 6, *alares nasi*)—to the eyes and eyelids, (7,
ophthalmic)—to the forehead, (8, *frontal*)—to the
eye sockets, (9, *upper orbital*; 10, *lower orbital*)—
to the temples, (11, *deep temporal*; 12, *outer temporal*; 13, *front temporal*). All these are still
farther sub-divided, and numerous other branches
are given off by the carotids.

Starting again from the great bend or arch of the
aorta (17), we follow, under the collar-bone, the *sub-
clavian* artery (22); farther on, in the shoulder, this
becomes the *axillary* artery (55); in the arm it is
called the *brachial* artery (60); in the forearm the
brachial artery divides to form the *radial* artery
(62) and the *ulnar* artery (65). These are the
arterial trunk lines of the arm, which, by their
numerous branches, supply every part of that limb,
down to the tips of the fingers, with blood.

As the aorta passes downward through the chest,
it is called the *thoracic aorta*, and gives off branches
to the pericardium, lungs, and trachea, oesophagus,

partition between the lungs, and the walls of the chest. After it passes through the diaphragm into the abdomen, it is called the *abdominal aorta* (34). Then it gives off the *phrenic* arteries (35) to supply the diaphragm, and walls of chest and abdomen. Next we observe the *coeliac axis* (36), the common origin of three important aortal branches, namely, the *gastric* artery passing to the stomach, the *hepatic* artery passing to the liver, and the *splenic* artery going to the spleen and other abdominal organs. Then the *superior mesenteric* artery (37) starts off to supply the upper part of the intestinal region. Immediately below the upper mesenteric artery, the *renal* arteries (41) branch off to the right and left kidneys. At (38) the *spermatic* arteries branch off to be distributed to the reproductive organs, the *inferior mesenteric* artery (39) supplies the lower inner abdominal parts. One of its main branches is the *haemorrhoidal* artery (40). All these aortal branches are much farther subdivided. Finally, in the abdomen, rather to the left side of the middle abdominal line, at a point corresponding within to the left side of the umbilicus (navel), the aorta divides into the two great common *iliac* arteries (42). After giving small branches to the peritoneum, loins, and other parts. the iliac arteries branch off again into two main channels, the *internal* and the *external* iliacs (43, 44) to supply with numerous branches the internal and external parts of the pelvic and hip regions. The continuation of the *ex-*

ternal iliac artery (44) into the thigh, becomes the *femoral* artery (70). In the upper part of the thigh the femoral artery runs rather close to the surface. Through the middle thigh it runs deeper in two branches, (70, 71). The figure shows a number of the larger branches of the femoral. In the lower part of the upper leg the femoral turns deeper or backward, throwing out, at the same time, what are called the three *perforating* arteries (83, 84, 85, right-hand figure). Above the knee, the femoral becomes the *popliteal* artery (86). Below the knee, the popliteal divides to form the *anterior tibial* (74, full figure, 91, lower leg figure), and *posterior tibial* (93). The *peroneal* artery (92), is a large branch of the posterior tibial. The two last named arteries send out their branches to all parts of the lower leg and foot.

The foregoing is a description of the main arterial channels which are distributed to the various parts of the body. The numerous small arterial branches which diverge from the great arteries just learned, by their farther division into innumerable capillary channels, furnish blood to every tissue at all points in the whole structure.

Having passed through the capillary meshes, and having delivered its nutrient substances for the growth and repair of the body, and taken up, instead, waste material and impurities from the body, the blood is re-collected by numerous veinlets which soon run together to form important venous channels. We have an illustration of this on the chart, in the right leg figure, where the veins of that limb which are near to the surface, are shown to form the smaller and the larger *saphenous* veins (96, 95). Similarly, in the right arm, the formation of the *cephalic* (57), basilic (58) and *median* vein (59) is shown. In the lower part of the trunk, the blood from the lower extremities and from the hips and lower abdominal parts is collected into the **great** *iliac* veins (51), corresponding to the great iliac arteries. These iliac veins unite to form the large *ascending vena cava* (47). Some of the larger tributaries flowing into the last named channel of the venous system are the *spermatic* veins (50), the *renal* veins (49), the *hepatic* veins (48), and the *phrenic* veins, corresponding to the phrenic arteries.

From above, on the right side, the *sub-clavian* (27) and the *jugular* veins unite to form the right *innominate* vein (24). On the left side, the *left sub-clavian*, the *jugular* (25) and the *thyroid* veins (28) unite to form the *left innominate* vein (23). Now the two innominate veins (23, 24) join to form the great *descending* vena cava (14), which enters the right auricle of the heart.

The *pulmonary* veins (31) are the exception to the rule that veins carry impure blood. They return the purified blood from the lungs to the heart.

The great veins from the abdominal organs, which unite to form the portal vein have been described under the portal circulation.

The manner in which the supply of **Regulation of Blood Supply.** blood to the various parts and organs of the body is regulated, is one of the most interesting subjects of physiological study. "Nature is never wasteful, whether of power or material. All her operations are performed on the principle of a proper economy." When any limb or organ of the body performs work, it suffers wear, which calls for a larger supply of repairing material than would be needed if it were more inactive. In general, the proper flow of blood to any part is in proportion to the activity of such part. How is this blood-flow adjusted?

The arteries have three distinct coats. **Vaso-motor Nerves and Muscles.** In the smaller arteries, the middle coat is almost entirely composed of tiny muscular fibers. These muscular fibers, like the fibers of all other muscles, are in nervous connection with controlling centers of the nervous system. By this plan of structure, the muscles of the small ramifying arteries are under involuntary or sympathetic nervous control. By their action the capacity of the small arterial channels is increased or diminished. The nerves which control these blood-vessel muscles are called *vaso-motor nerves*, and their regulation of blood-supply is called *vaso-motor action*. So "vaso-motor nerves" may be understood to mean "blood-vessel controlling nerves." Now, for illustration, when the brain is actively exercised, as in hard study, the muscles of the blood-vessels of the brain, by sympathetic action of the vaso-motor nerves which control them, are made to act in such a way as to increase the dimension of these blood-vessels, and thus increase the flow of blood to the brain. When the brain returns to a state of rest, a reverse action of these tiny muscles restores the blood-vessels to their former diminished capacity.

So, while the general circulation ordinarily goes on at a uniform rate, the local supply of blood is subject to much variation. This arrangement—as wise as it is wonderful—has been nicely illustrated as follows: "If when one group of muscles was set at work and needed an extra blood supply, this could be attained only by increasing the heart's activity and keeping up a faster blood-flow everywhere through the body, there would be a clear waste of force, much as if the chandeliers in a house were so arranged that when a large flame was wanted at one burner, it could only be obtained by turning more gas on at all the rest. Besides the big tap at the gas-meter, regulating the *general* supply of the house, *local* taps at each burner are required, which regulate the gas supply to each flame, independently of the rest. Vaso-motor action is a similar arrangement in the blood-supply of the body."

A GENERAL VIEW OF THE COMBINATION AND RE-
LATION OF THE VARIOUS SYSTEMS OF THE
BODY, AS SHOWN BY THE MANIKINS
OF THE ANATOMICAL AID.

MANIKIN OF THE HEAD.

First Section. This first section of the manikin
of the head represents as nearly as possible the perfection of external form and feature. The variations
from such an ideal figure, which give such an end-
less variety to the personal appearance of men and
women, are chiefly due to the following physiological
conditions: 1.—Differences in shape and prominence
of the bones of the skull and face. 2.—Peculiarities
in the size, shape and position of the ears. 3.—Shape
and prominence of the bones and cartilages of the nose.
4.—Differences in plumpness of muscle and fat
underlying the skin. 5.—Variable shapes and full-
ness of the lips. 6.—Protruding or retired eyes.
7.—Color of eyes and hair. 8.—Complexion of the
skin. 9.—*Expression* of the eye. 10.—The *muscular habits* of the countenance.

Second Section. Close under the skin and its
underlying cushions of fat, appear vessels of the
circulatory system and numerous *muscular* bands.
Of the former system, the *external carotid* artery
(24), sends the *posterior auricular* to distribute
blood to the region back of the ears; the outer *temporal* (26) to the temples; the *transverse facial* (27)
to the middle portions of the face; the *external*

maxillary (28) to the jaws; the *lower labial* (29) to the lower lip; the *angular* (30) to the nose, and the *frontal* (31) to the forehead region. The *frontal* vein (23) from the forehead, the *ophthalmic* (22) from the eye, the *temporal* (21) from the temples, the *labial* (20) from the lips, and the *occipital* (17) from the back part of the head, with others, unite in the outer and inner *jugular veins* (16, 19) to drain the impure blood from the head towards the heart.

Of the muscles, the *sterno-cleido-mastoid*, the prominent muscle on the side of the neck, starts from the sternum and clavicle bones and goes obliquely upward and is attached above to the *mastoid process* of the temporal bone. There are, of course, two of these muscles—one on each side. When one acts alone, it brings the head obliquely forward. When both act at the same time, they bring the head directly forward.

Just back of the muscle just described, lies the *splenius* muscle (2). Starts from the five upper vertebral bones, and goes to the temporal bone. It acts counter to the sterno-cleido-mastoid, since it aids in moving the head backward.

The *occipito-frontal* muscle (3) passes from the back part of the head, over the top, to the lower edge of the forehead. It raises the eye-brows and wrinkles the forehead horizontally.

The *masseter*, or chewing muscle (5), passes from the malar bone (cheek-bone) and the upper maxillary (upper jaw-bone) above, to the lower jaw-bone.

The *buccinator*, or blowing muscle (6), forms the principal part of the cheek. It passes forward to the corners of the mouth. When it contracts, it draws the angle of the mouth backward. The trumpeter or cornetist makes good use of this muscle.

The two zygomatic muscles—*zygomatic major* (7) and *zygomatic minor* (8)—pass from the cheek-bone to the upper lip, which they raise and draw outward as in laughing.

The circular muscle about the *mouth—orbicularis oris* (9)—in ordinary contraction, closes the lips; in strong contraction, it puckers the mouth.

The *depressor anguli oris* (10) starts from the lower edge of the lower jaw and passes up to the corner of the mouth, which, by its contraction, is drawn downward.

The *depressor labii inferioris* (11) pulls down the lower lip.

The circular muscle about the eye—*orbicularis palpebrarum* (12)— closes the eye.

The *levator alœ nasi* (14) contracts to enlarge the nostrils.

Third Section. Here the bony framework of the head is brought to view. The *frontal* bone (32) giving shape to the forehead, the *parietals* (33), the upper sides of the skull; the *temporal* (34), the lower skull sides, or temples. Two important projections or processes of the temporal bone are shown: The *zygomatic process* (35) which goes forward

to meet a similar process of the *malar* or cheek-bone
(38) to form the *zygomatic arch*, under which several muscles pass, and the *mastoid process*, which
may be felt as quite a prominence direc ly behind
the ears. Parts of the *nasal, upper maxillary* and
lower maxillary appear at (39, 40, 41). The *suture*
joint connection between the bones of the head is
also shown. The outer layer of a part of the bones
at the back of the skull is represented as removed.
to show the interior spongy texture between the two
bony layers.

Of the deeper muscles which here appear, we have
another view of the *orbicularis oris* (53) and the
buccinator (54).

The two pterygoid muscles (55, 56) are very prominently concerned in the process of chewing.

The *stylo-hyoid* muscle starts from the styloid
process of the temporal bone and passes down to the
bone at the base of the tongue. It draws this
tongue-bone upward and backward.

The remaining muscles shown on this section
perform various movements of the pharynx, tongue,
head and neck.

Fourth Section. Here the left side of the skull
is removed, giving us a view into the very interior
of the brain—the great capital of the nervous system.

The delicate coverings of the brain are shown—
(a a) being the outer and harder one, called the
dura mater. At (u) and (v), portions of the outer
surface of both the cerebrum and cerebellum—

larger and smaller brains—are shown, with some of the veins which collect the blood from them. Some of the nerves passing from the brain to the eye and various parts of the face also appear. The spinal column is shown as cut vertically in two, revealing the spinal cord (w) with its branching nerves, and the spinous processes of the vertebræ with the binding ligaments between them (89).

Fifth Section. This brings us to the very middle of the skull and its contents. We are looking at the inner surface of the right hemisphere of the brain. Some of the numerous arteries which furnish it richly with blood, are shown. The *corpus callosum* (127), is a peculiar white body, comparatively hard, at the base of the cerebrum. The interesting tree-like structure, called the *arbor vitae* (123), which appears when the *cerebellum* is vertically bisected, is nicely represented. At (124) is the upper part of the spinal cord (*medulla oblongata*), and at (150) the spinal cord itself is shown to pass down through its vertebral canal.

MANIKIN OF THE BODY.

On removing from the trunk its integument, or skin covering, the muscular system at once comes prominently into view, showing the great muscles which pass from the head, through the neck, to the chest and to the shoulders; those which pass from the chest to the shoulder and arm; the outer muscles of the abdomen, and the large muscles which pass from the trunk to the thigh.

The great surface muscles of the trunk being removed, the bony frame-work of the chest appears in the *sternum* or breast-bone (2), to which the twelve pairs of ribs (3, 4), are attached by their cartilages (5). Within these the muscular system is again represented by the *intercostal* muscles (6). Within these, again lies the *pleura* shown on the opposite side.

The ribs, with their connecting muscles and lining pleura being removed, the organs of the chest—lungs and heart—are seen in place. First we notice, the light, spongy or cellular structure of the lungs; also their lobular divisions—three on the right and two on the left side.

On the next section the circulatory system is prominently represented, first, by its great central organ, the heart (10), lying beneath and between the right and left lungs. Its auricle and ventricle chambers lie within, as located by (13, 14, 11, 12). The great pulmonary artery (16) carries the impure blood from the right ventricle to the lungs. The pulmonary veins (18) bring the purified blood back from the lungs to the left auricle of the heart. The *descending vena cava* (17) formed by the large veins which drain the head, neck, arms, and upper chest, empties the impure blood into the right auricle of the heart. From the left ventricle of the heart, the great *aorta* (15) springs to deliver pure blood to numerous arteries throughout the body.

The next section is chiefly illustrative of the respiratory system, showing, by a special manikin, the structure of the organ of the voice—the *larynx*— at the top of the *trachea* (24) or wind-pipe. We observe the division of the trachea into the *bronchial tubes* (25), and the farther subdivision of these into numerous air tubes which ramify the lungs and terminate in countless air-cells. At (f), (g,) and (c), respectively, are portions of the pulmonary artery, aorta and descending vena cava. The heart chambers are laid open and the inner valves appear.

In the abdominal region, after the skin and muscular wall is removed, the *peritoneum* (a) appears. This forms a lining of the abdomen, and, at the same time, by its numerous folds, or *omenta* (5), covers the abdominal organs.

Removing the peritoneal veil, the digestive system is brought to view. First, the middle portion of the small intestines—the *jejunum* (a), and the latter portion—the *ilium* (b); then, on the next section, the divisions of the large intestine, as, the *transverse colon* (a), the *descending colon* (d), the beginning of the *ascending colon* (f). At (h) is the end of the *duodenum;* at (i) the beginning of the *jejunum;* at (k) the end of the *ilium.* The mesenteric folds of the peritoneum, winding between and around the intestines, are shown at (g).

At this point we notice the diaphragm (26)—the muscular partition between the chest and the abdomen. Observing its relation to the abdominal organs, we can see why it is that when the diaphragm presses these organs down, as it does in the act of inspiration, the abdominal wall must expand and move outward.

Directly under the diaphragm partition we meet the liver, lying over and to the right of the stomach. It has two lobes—called the right and left. Its inner section shows its importance as a blood-receiving and blood-changing organ. At (g) is the bile reservoir or *gall-bladder.*

Next we observe the position and shape of the stomach (31). We see how it connects with the duodenal end (a) of the small intestines. In this portion of the small intestines the process of digestion is completed. Where the stomach ends and the duodenum begins, is the *pylorus* valve. The *oesophagus* (35) is the tube leading from the throat to the stomach, through which food passes in swallowing.

The next section shows the inner wall of the stomach. Its rosy tint comes from the great number of blood-vessels which run through the inner coat. The pylorus is situated at (c). The blood-vessels which ramify the intestinal wall are shown.

We notice next the position and structure of the *kidneys* (29). The inner section (*right kidney*) shows how blood is brought to these organs by the *renal arteries* (g), and carried away from them by the *renal veins* (55). The inner section (*left kidney*) shows the *pyramids* of *Malpighii*, in which the urine is secreted from the blood. When thus secreted, the urine trickles into the kidney reservoir—*pelvis renalis* (c)—from whence it is carried through the *ureters* (56) into the *bladder* (57) and from thence out of the body.

The *pancreas* (59) lies behind the stomach and left kidney. It secretes the pancreatic fluid which serves in the process of digestion. The *spleen* (60) lies on the left side under the left end of the stomach. The use of this organ is not understood.

On the last section of the body manikin, the bony system is represented by sectional views of the *clavicle*, or collar-bone (48), the *scapula*, or shoulder-blade (49), the upper part of the *humerus*, or arm-bone (50), the *ribs* (53), the *ossa innominata*, or pelvic bones (58), and the head (a) and larger process (b) of the *femur*, or thigh bone (61).

The muscular system is here represented by the deep muscles of the neck (47), the *deltoid* and *brachial* muscles of the shoulder (51, 52), the *intercostal* muscles between the ribs, and the *psoas* (54) and other muscles of the loins and thighs.

Observe the *thoracic duct* (46) as it begins in the *chyle receptacle* (46 a) in the abdomen, and ascends behind the oesophagus to the left shoulder region (46 b), where it empties chyle and lymph into the left subclavian vein (38).

The circulatory system appears in the subclavian arteries and veins (39, 37, 38), the *thoracic aorta* (45), the *abdominal aorta* (42), and the *iliac* divisions of the aorta (41, 44), the *iliac* veins (43) and their union into the great *ascending vena cava* (40).

OUTLINE LESSONS

SUGGESTING A METHOD OF ELEMENTARY DRILL.

EXERCISE I.

FIRST SKELETON PLATE.

Teacher.—Of what is this a picture, class?

George perhaps answers: "A man." Fannie may say: "They are bones." After drawing out as many ideas as possible concerning the figure on the plate, state as follows:

"This is a picture of the framework of the body, and it is called a **skeleton**. All pronounce the word."

Pupils.—Skeleton.

T.—The framework or skeleton of the body contains 208 separate pieces called **bones**. Now let us notice the different shapes of these bones. Charles, what particular one are you noticing now?

Charles.—The long one which forms the leg.

T.—Yes, that is a long, straight bone. Hattie, what different shaped one do you see?

Hattie.—I am looking at the flat one at the shoulder.

T.—Yes, notice that in the lower part of the body, there are also some large flat bones (3). What other shapes of bones do you see, class?

Some will notice the broad curved bones of the skull, others the slender, bent bones of the ribs, others the small round bones of the wrist, and so forth. In this way, drill into the minds of your pupils the idea of **shape** as a property of bones.

T.—Now since there is such a large number of bones in the body, it will be best for us to study them in sets or groups. How many different groups of bones do there seem to be in the skeleton?

Perhaps half-a-dozen hands will go up. One says he sees **two** groups; another **three,** some, possibly **four.** After letting them state their various observations on this point, state as follows:

"There are four sets or divisions of the bones of the body—the bones of the head, the bones of the trunk, the bones of the upper limbs or extremities, and the bones of the lower extremities."

This will be sufficient material for the first exercise.

T.—How many of you will be willing to bring a specimen of a bone to the class to-morrow, when we have our next drill?

A number of hands will go up. Appoint Fred to bring a long slender bone; Edward a flat bone, and so on; so that you may have the differently shaped bones for next day's review of the first lesson.

T.—Now let us see what we have learned. What is the framework of the body called?

P.—The skeleton.

T.—Of what is it composed?

P.—Of bones.

T.—How many bones in the skeleton?

'P.—Two hundred and eight.

T.—What shapes of bones have we found?

P.—Long, straight bones; flat bones; curved bones and round bones.

T.—How many divisions or groups of bones in the skeleton?

P.—Four.

T.—Name them?

P.—The head, the trunk, the upper extremities and the lower extremities.

T.—In to-morrow's lesson we will learn all we can about the bones of the head.

EXERCISE II.

FIRST SKELETON PLATE.

T.—Now let me see how much you remember of your first physiology lesson. How many sets or groups of bones did we find in the skeleton?

P.—Four.

T.—(Pointing to skeleton.) Herbert, what is this group called?

Herbert.—The head.

T.—Ella, will you name this division of the skeleton?

Ella.—The trunk.

T.—Jennie, what are these parts called?

Jennie.—The upper limbs or extremities.

T.—Have you brought in the specimens which you promised me yesterday?

(Specimens are handed in and carefully compared and examined.)

T.—Now to-day, we will study the bones of the

head. Draw out ideas about the shape. Encourage the pupils in expressing their ideas by comparison. Even if Harry should compare the head, in shape, to his foot-ball, do not ridicule but encourage the observation.

T.—(*Pointing to the forehead on the chart.*)   What do you call this part of the head?

Some will answer: " The front part."   Others may say: " The forehead."   Approve both answers.

T.—Place your finger on the front part of your head.   Do you feel a bone there?   ·

P.—Yes.

T.—This large bone (1) in the front part of the head is called the **frontal** bone.

(Teacher spells the name and writes it on the blackboard.)

T.—What is this part of the head called?

P.—The side.

T.—Yes; the upper sides of the head are formed by the **parietal** bones (2).   (Spell the name and write on blackboard.)   Here (12) the dividing line between the frontal bone and the parietal bone is very clearly shown. On the second or back-view chart of the skeleton (*turn to it*) the two parietal bones (2) are more fully shown.

(*Turn back to the first chart.*)

T.—The lower sides of the head are called the temples.   Place your finger on the side of your head, just in front of the upper part of your ear.   The bone which you feel is this one numbered (3) on the chart.   It is called the **temporal** bone.   We notice this dividing line between this temporal bone and the frontal and parietal bones.   The back-view of the head (*turn to*

next chart) shows the back part ot the temporal bone.
Of course, there are two temporal bones—one on each
lower side of the head.

T.—Touch the back part of your head. Now you
feel the bone which is shown here (4). Its name may
be hard for you to remember. I will write it, as I
spell it, on the board: O-c-c-i-p-i-t-a-l, occipital. What
is this called?

P.—Occipital.

T.—Now how many and what bones of the head
have we learned about?

P.—One frontal bone, two parietal bones, two temporal bones, and one occipital bone.

T.—Right, so far; besides these there are two more,
which form, as it were, the floor or bottom of the bone-
box which forms the upper part of the head. These
two bones which cannot be shown on the chart because
they are so much concealed within the other bones of
the head, are called the **sphenoid** bone and the **ethmoid** bone. Now these two bones, with those we have
learned about before, make how many?

P.—Eight.

T.—Right; these eight bones of the head are put
together in such a box-like form, to contain and protect
one of the most delicate and important parts of the
body. How many of you know what is contained in
this box? (Hands up.) All who know may tell me.

P.—The brain.

T.—Correct. We shall learn about the brain hereafter. Now the upper part of the head, which we
have called the brain-box, formed by the eight bones
of which we have learned, is called the **skull.** Will

you now name the eight bones which form the skull?

P.—One frontal, two parietal, two temporal, one occipital, one sphenoid and one ethmoid bone.

T.—Very good. What bone forms the front of the skull or brain-box?

P.—The frontal bone.

T.—What bones meet above to form the top of the skull?

P.—The frontal and the parietal bones.

T.—What bones form the sides of the skull?

P.—The parietal and the temporal bones.

T.—What bone forms the back of the skull?

P.—The occipital bone.

T.—What bones form the floor or bottom of the skull?

P.—The sphenoid and the ethmoid bones.

EXERCISE III.

FIRST SKELETON PLATE.

Note.—As these exercises are simply suggestive, the teacher must feel at liberty to vary them according to his pleasure. He must also use his judgment as to whether it is best for him to undertake to teach the pupils the names of the bones in these elementary drill lessons. Teachers who prefer to do so will find the names in the index to the skeleton charts.

T.—Thirty bones make up the whole skeleton of the head. How many of these did we find in the skull?

P.—Eight.

T.—There are four very small bones in each ear. The remaining bones of the head form the face. Ralph, will you tell me how many bones there must be in the face?

Ralph.—Fourteen.

T.—Very well. What different parts of the face do you notice?

P.—Nose, mouth, cheek, chin, eyes.

T.—What is this (6)?

P.—The nose.

T.—Is it the whole of the nose?

P.—No; it is only the bony part of the nose.

T.—Yes; it is called the bridge of the nose, and is formed by two small bones. (*Nasal bones.*) What is this (4) part of the face called?

P.—The cheek.

T.—You can easily feel your cheek bone (*malar*) on each side of your face. Laura, what would you call this (7) bone?

Laura.—Jaw-bone.

T.—Robert, what would you call this (8) bone?

Robert.—Jaw-bone.

T.—You are both right; but how shall we tell these jaw-bones apart?

P.—By calling them upper and lower jaw-bones.

T.—Do you notice anything more about these jaw-bones?

Ralph.—The teeth seem to be fastened into them.

T.—So they are. The teeth are set very firmly into pits or sockets of these jaw-bones. Besides these two nose-bones, two cheek-bones, and two jaw-bones, there are eight other face bones, which are more concealed, and whose names you will learn by and by. Now what seems to be the use of these face bones? (Draw out: to shape the nose, the cheek, the chin, and to contain the mouth, teeth, eyes, and so forth.)

T.—Now let me see what you know about the bones of the head.   What is this (1) bone called?

P.—The frontal bone.

T.—This bone (2)?

P.—Parietal bone.

T.—This on the lower side of the head (3)?

P.—The temporal bone.

T.—And what bone forms the back of the head?

P.—The occipital bone.

T.—Besides these there are how many more skull bones?

P.—Two; the sphenoid and the ethmoid.

T.—How many skull bones in all?

P.—Eight.

T.—What is the use of the skull?

P.—To enclose and protect the brain.

T.—How many bones in each ear?

P.—Four.

T.—How many bones in the face?

P.—Fourteen.

T.—What face bones can you mention?

P.—Two nose bones; two cheek bones; two jaw-bones.

T.—How many more are added to these six to form the face?

P.—Eight.

T.—What are set in the jaw-bones?

P.—The teeth.

T.—In our next lesson we will learn about the bones of the trunk.

EXERCISE IV.

FIRST SKELETON CHART.

T.—What is this division of the skeleton called?

P.—The trunk.

T.—That seems to be a very good name for it. You know, a trunk is used to contain and protect such articles as are put into it. So here, the trunk of the body, especially as is shown by the shape of its upper part, is intended to contain some important inside parts of the body. Can any one of you name some part of the body which is contained in this trunk?

Mattie.—I think the heart is in the trunk.

Albert.—The lungs.

T.—The upper part of the trunk which encloses the heart and the lungs, is called the **chest.**

Charles.—That is also the name of something which shuts in things for safe-keeping.

T.—You are right, Charles. The lower part of the trunk is called the **abdomen.** Can you tell me any parts of the body which you think are in the abdomen?

Harry.—The stomach.

Nellie.—The liver.

T.—Correct. Now let us see how the framework of the trunk is put together. Notice this long pillar of bones at the back. What is this called?

P.—The back-bone.

T.—Yes; it is also called the spine or spinal column. The next chart (*turn to it*) gives us a very good view of the spinal column. John, what do you notice at its upper end?

John.—The head rests on it.

T.—Right; Ada, what do you see at the lower part of it?

Ada.—There are large bones fastened to it.

T.—Hattie, what do you notice about its middle part?

Hattie.—There are many bones fastened to it.

T.—This great spinal pillar, or backbone, is built up of twenty-six separate bones. Please count them aloud, as I point to them from the top down.

(Teacher point to each, from (1) at the top, to (5) at the bottom.)

T.—There must be some good reason why the backbone is composed of so many parts instead of being but a single bone. Thomas, can you see any reason for it?

Thomas.—I think it makes the back stronger.

T.—You are right, Thomas. What more does the class think about it?

Ida.—We could not bend the back if it were all one bone.

T.—That is true. Each one of these bones is called a **vertebra**. This name is spelled, v-e-r-t-e-b-r-a. Albert, please write it on the board. Now, please notice that these vertebrae do not rest directly upon each other. We see white, cushion-like bodies placed between them. (*See first chart, between lower vertebræ.*) These are composed of tough material, but softer than bones. This material is called **cartilage**. It makes the backbone springy: that is, it makes it easier to bend the back, and also prevents injury to the brain from heavy jars to the body.

EXERCISE V.

SECOND SKELETON CHART.

T.—What is this (14) great pillar of the body called?

P.—The backbone, or spinal column.

T.—Of how many bones is it composed?

P.—Twenty-six.

T.—What is each bone called?

P.—A vertebra.

T.—What are placed between the vertebræ of the backbone?

P.—Cushions of cartilage.

T.—For what purpose?

P.—To enable us to bend the body, and to soften the effects on the brain, of heavy footsteps or violent jars of the body.

T.—Now let us next learn the four divisions of this spinal column. (*Pointing to upper vertebræ.*) What is this part of the body called?

P.—The neck.

T.—The neck is formed by a number of vertebræ of the back-bone. Let us see how many. As I point to each, please count aloud. (Pupils count seven.) How many vertebræ are in the neck?

P.—Seven.

T.—Can you name these long curved bones?

P.—Ribs.

T.—Right. They remind us of barrel-hoops. We see that they are in pairs. Let us count these pairs of ribs. (*Point to each from top down while pupils count.*) Twelve pairs. To what are these ribs fastened behind?

P.—To the vertebræ of the backbone.

T.—So how many vertebræ of the back have ribs attached to them?

P.—Twelve.

T.—These twelve vertebræ which have ribs attached to them are called vertebræ of the back. Below these are the five large vertebræ of the loins. (*Point to each and count.*) Now what three divisions of the backbone have we found?

P.—Seven vertebræ of the neck, twelve vertebræ of the back, and five vertebræ of the loins.

T.—Now the next backbone is the **sacrum** (4). Notice how it is wedged in between these (1) two large hip bones. Below the sacrum is a small end-bone called the **coccyx** (5). Do you notice any difference in the ribs?

Harry.—Some are loose at one end.

T.—Are all of them attached behind?

P.—They are.

(*Turn back to first skeleton plate.*)

T.—What part of the body does this flat bone (9, 10) occupy?

P.—The breast.

T.—So it is called the breast-bone; but its more proper name is the **sternum**. Let us see how many pairs of ribs are joined to the breast-bone. Count as I point to them. How many?

P.—Seven pairs.

T.—These seven pairs are called the **true ribs.** The other five pairs are not directly united to the breast-bone; but they are united to each other and to the seventh true rib. These five pairs are called **false**

ribs. Notice that the true ribs seem to be united to the breast-bone by a separate piece of material. This separate piece is cartilage. Since the ribs are united to the sternum by this springy piece of cartilage, there is much greater protection to the delicate organs inside, since a blow on the chest is very much broken or softened by the springy nature of the chest wall. Besides, this arrangement gives the chest a much easier motion in breathing, as we shall learn hereafter. Have we spoken of all the bones of the trunk?

Carrie.—What are those large bones at the lower part of the trunk?

T.—These (3) large bones make a wide bowl-shaped support for the contents of the abdomen, and a firm place of attachment for the bones of the lower limbs. You may remember them for the present as the hip-bones. Now let us see what you know about the trunk. How many bones in the trunk?

P.—Fifty-three.

T.—What important organs are contained in the trunk?

P.—Heart, lungs, stomach, liver, and others.

T.—What is the upper part of the trunk called?

P.—The chest.

T.—The lower part?

P.—The abdomen.

T.—On what strong pillar of bones does the head rest?

P.—On the spinal column, or backbone.

T.—How many bones in the spinal column?

P.—Twenty-six.

T.—How are they united?

P.—There are cushions of softer material placed between them.

T.—What is each one of these bones called?

P.—A vertebra.

T.—What can you say about the number of ribs?

Frank.—There are twelve pairs; seven pairs are true ribs attached in front to the breast-bone; five pairs are false ribs, not directly attached to the breast-bone.

T.—Why should these bones (3) be so large?

P.—To support the organs in the abdomen and to give a firm attachment to the bones of the lower limbs.

EXERCISE VI.

BACK VIEW OF THE SKELETON.

T.—What is this part to which the arms are attached called?

P.—The shoulder.

T.—What do you notice about this (13) bone?

P.—It is very broad and flat.

T.—It is commonly called the shoulder-olade.

(Turn to first skeleton chart.)

T.—This bone (8) is called the collar-bone. I give you the common names now; at the proper time, hereafter, you will learn their scientific names. The collar-bone rests, as we see, at one end against the breast-bone and at the other against the shoulder-bone. Have you any idea what the use of this bone may be?

David.—It keeps the shoulders braced apart.

Ralph.—I should think it would keep the shoulders from resting too heavily on the ribs of the chest.

T.—You are both right. All grasp firmly the bone

of your upper arm. This (1) is the bone you feel. This large bone is called the **humerus.** What is this joint at its lower end?

P.—The elbow.

T.—How many bones are in the fore-arm?

P.—Two (2, 3).

T.—What is this part (4) of the arm called?

P.—The wrist.

T.—Robert, will you step to the chart and count the small bones which form the wrist? (Robert counts eight.) Now count the bones in the middle or palm of the hand as they are shown on the left hand, on the chart. (Robert counts five.) Now count the bones in the fingers. (Robert counts two in the thumb and three in each finger—fourteen in all.)

T.—Now we will notice how these bones are bound to each other. They seem to be held together by strong cords, as we see here at this wrist (*left arm*), between the armbones, at the elbow, and at the shoulder. These strong bone-binding cords are called **ligaments.** They are so strong that sometimes the bones themselves will break easier than the ligaments which bind them together. Now let us review the lesson. What is this bone called (16)?

P.—The shoulder-blade.

T.—Why should it be so strong?

P.—Because it supports the arms.

T.—Name this (8) bone.

P.—The collar-bone.

(Draw out, by questions, its use.)

T.—Do you remember the name of this bone?

P.—The humerus.

T.—How many bones in the **fore-arm?**

P.—Two.

·T.—In the wrist?

P.—Eight.

T.—In the middle hand?

P.—Five.

T.—In the fingers of each hand?

P.—Fourteen.

T.—Now who will tell me first how many bones in collar, shoulder, arm and hand?

Harry.—Thirty-two.

T.—A very ready answer. So how many bones in both of the upper extremities?

P.--Sixty-four.

EXERCISE VII.

FIRST SKELETON CHART.

Now we will study the lower extremities. What do you notice about this bone?

Maggie.—It is the largest bone in the body.

T.—It is both the largest and the strongest bone in the body. It is called the **femur.** To what is it connected above?

P.—To the large hip-bone.

T.—So this (2) is called the hip-joint. What is the joint at the lower end of the femur called?

P.—The knee-joint.

T.—As we see here (XIV, XV, XVI,) the large leg-bone is bound to the hip-bone by the strongest ligaments of the body. Lillie, what do you notice about this knee-joint?

Lillie.—There seems to be an extra little bone there.

T.—Yes; this round, flat bone (2) is called the knee-pan. It protects the knee-joint. How many bones in the lower leg?

P —Two.

T.—What difference do you see in them?

P.—One is a very strong bone, and the other is quite slender.

T.—What is the back part of the foot called?

P.—The heel.

T.—What part is just in front of the heel?

P.—The instep.

T.—There are seven bones in the heel and instep, five in the middle, and fourteen in the toes, of each foot. Edith, will you tell us how many bones there are in each lower limb.

Edith.—Thirty.

T.—You have counted correctly. That would make how many bones in both of the lower extremities?

Edith.—Sixty.

T.—Now, class, what are the four divisions of the skeleton?

P.—The head, the trunk, the upper limbs and the lower limbs.

T.—How many bones did we find in the head?

P.—Thirty.

T.—How many in the trunk?

P.—Fifty-three.

T.—How many in the upper extremities?

P.—Sixty-four.

T.—How many in the lower extremities?

P.—Sixty.

T.—What is the sum of 30+53+64+60?

P.—Two hundred and seven.

T.—There is one more extra bone at the root of the tongue. This would make how many in the whole body?

P.—Two hundred and eight.

EXERCISE VIII.

CHART OF MUSCULAR SYSTEM.

T.—We will now learn about the **muscles** of the body. What do you notice about this figure of the body which is now before us?

James.—It looks very different from the skeleton.

Addie.—It looks more like the real body.

Ethel.—The bones are nearly all covered up.

T.—Your answers are all quite good. These red colored parts on this chart represent muscles of the body. What does their appearance remind you of?

Fannie.—They look like raw lean meat.

T.—And that is what the muscles of the body are, Fannie. The muscles form the raw, lean flesh of the body. They are made of many strings or fibers, something like a skein of yarn. You can see these strands or muscle-fibers in a piece of beef that has been well cooked. Such a piece of meat may be pulled into string-like strands. The next time you have some cold cooked beef on your table, at home, I wish you to separate a small piece into its muscular fibers.

Charles.—Are all muscles red?

T.—Muscle of beef, as we get it from the market, is

quite red. The muscle of our bodies is also red, but not quite so red as raw beef. Muscle of pork, as you know, is of a much paler red. Harry, what have you noticed about the meat on the breast bone of a chicken, at your dinner table?

Harry.—It is nearly white.

T.—So not all muscle is red. But it is generally red.

William.—What is the use of the muscles?

T.—That is a very proper question, for which we are just ready. The use of the bones, as we have learned, is to form the framework of the body The use of the muscles is to move parts of the body, or the whole of the body. For this purpose the muscles are attached to the bones.

Fred.—How can the muscles move the bones?

T.—I will try to explain it to you. The threads or fibers of each muscle have the power of shortening up lengthwise and swelling out sideways. What effect would that have on the whole muscle?

Fred.—It would shorten and thicken the muscle.

T.—Exactly. Now, if such a muscle is fastened at its two ends to different bones, what will happen when the muscle shortens up?

Edith.—It will move the bones.

T.—Just so; all the movements of the body, no matter how vigorous or how slight they may be, are produced by such a shortening of muscles. This shorten-ing of a muscle, in pulling or moving any part of the body, is called **contraction.**

Albert.—But how does a bone get back to its first place ?

T.—The muscle which has contracted or shortened

to move it, must now lengthen out to **let** the bone back ;
at the same time some other muscle or muscles must
contract or shorten to **draw** the bone back. Do you
understand this, Albert?

Albert.—I do.

T.—The lengthening of a muscle to let a bone, or
other part which it has moved back, to its former position, is called **relaxation.** Now let us gather up
what we have so far learned about the muscles. What
part of the body do the muscles form?

P.—The lean flesh of the body.

T.—Of what is a muscle composed?

P.—Of many fine threads or fibers.

T.—What is the color of the muscles of our bodies?

P.—Red.

T.—Are the muscles of all animals red?

P.—They are not. But the muscles are generally
red.

T.—What is the work of the muscles?

P.—To move the whole body, or one or more of its
parts.

T.—How do the muscles produce motion?

P.—By shortening up lengthwise and swelling out
sideways they pull on the parts to which they are
attached.

T.—What is the shortening up of a muscle called?

P.—Contraction.

T.—What is the return of the muscle to its usual
length called?

P.—Relaxation.

EXERCISE IX.

CHART OF MUSCULAR SYSTEM.

T.—In our last lesson we learned of what muscles are composed, what their work is, and how they act to move any part of the body. We may now get a still better idea of the great work which our muscles perform by studying a few of them. There are 527 muscles in your body. Let us look at this (1) one. Only its front part is shown on the chart. It is fastened behind to the occipital bone. It passes over the top of the head forward, to the skin of the forehead. Now, being fastened to that firm bone behind, and in front to the skin of the forehead and eyebrows, what effect do you think its contraction or shortening would produce?

Jennie.—I think it would move the skin on the forehead.

Lizzie.—I think it would raise the eyebrows.

T.—You are both right. Let us make this muscle show us how it acts. Look at my head while I make this muscle contract. What do you see?

P.—It wrinkles your forehead and raises your eyebrows.

T.—That is precisely the work of the muscle. The names of most muscles are quite long and too hard for you to learn now. They are generally named according to the bones which they connect, or their shape, or the kind of motion which they produce. I will give you this one name of this (1) first muscle which we are studying, as a specimen. It is called the **occipito-frontalis** muscle. I will write this name on the board. What do you notice about this name?

Grace.—It is a double name.

T.—So it is. Can any of you see why this muscle was probably called the **occipito-frontalis** muscle? (*Hands up*). Well, George, what do you say?

George.—Because it goes from the occipital bone to the front part of the head.

T.—Very well; that accounts for its name, exactly. What do you observe about this (15) muscle?

Ralph.—It is nearly circular and passes quite around the mouth.

T.—What do you suppose is the effect of its contraction?

Ralph.—I should think it would close the mouth.

T.—You are right. When it contracts gently, it closes the mouth, and when it contracts strongly, it puckers the lips. (*Illustrate.*) Do you see any other muscle of the same shape anywhere?

Laura.—There is one like it around the eye.

T.—What must be its use?

Laura.—To close the eyes?

T.—Yes; it closes the eyes.

Hold your left arm out straight. Now grasp it firmly with your right hand a little above the elbow. Now draw your fore-arm up towards your shoulder. What do you feel?

Robert.—I feel my arm swell out where I hold it.

(Let all the members of the class be sure that they feel the action of the muscle.)

T.—Frank, what have we learned to be the effect on a muscle when it swells out sideways?

Frank.—It gets shorter.

T.—Right; and getting shorter, what does this muscle which you feel do?

Frank.—It brings the arm up towards the shoulder.

T.—(*Pointing to (34) upper arm.*) This is the muscle whose action you have felt. So the work of this muscle is what?

P.—To draw the fore-arm towards the shoulder.

T.—Bring your arm towards the shoulder again. Grasp it very firmly, as before. Now slowly straighten the arm. What do you feel?

Ralph.—I feel the swelling of a muscle on the back of my arm.

T.—That was the action of this (36) muscle on the back part of the arm. While it **contracts,** the muscle of which we spoke before **relaxes.** So by this action of these two muscles, the fore-arm is brought back into a straight position.

This will be sufficient to show you how the muscles work. Just notice, before we close this exercise, these many strong muscles of the shoulder, trunk and lower limbs. Let me show you the longest muscle in the whole body. Here (60) it is. It is commonly called the **tailor muscle** because its use is to cross the legs, as tailors are accustomed to do.

EXERCISE X.

It is suggested to the teacher that this exercise be devoted to a familiar conversational lesson on the proper care and use of the muscles. Impress the pupils with the truth that while they are learning many things about the structure and use of the different parts of the body, it is no less important to learn how properly to train, and how to take good care of these organs. Impress them with the fact that our health, and consequently much of our comfort and happiness, and even our lives depend upon such care. Teach them that every organ of the body is liable to injury from misuse—that is, from either too much use or too little use. Explain to them what habits and acts will deform and weaken the bony framework of the body, especially in the years of childhood and early youth, when the whole bony system is comparatively tender and flexible. Explain how bad positions in sitting, standing or walking will permanently misshape the body.

So of the muscular system. Show the pupils, by familiar illustrative references, what plumpness and vigor proper exercise gives to the muscles. The well developed arm of the blacksmith will afford you a good illustration. On the other hand, caution your pupils against all forms of exercise which are too violent, or which tax severely only a few muscles. That kind of play, exercise or labor is most healthful which calls into use the moderate action of the greatest number of muscles. The strong muscles of the robust laborer come from a general use of his limbs and his trunk.

For farther suggestions concerning this lesson on the hygiene of the muscles, see pages 25 and 26 of this book.

EXERCISE XI.

CHART OF THE NERVOUS SYSTEM.

T.—We now come to a very interesting part of the study of the body. We have learned about the bony framework of the skeleton. We have also learned about the muscles which cover the skeleton and move its different parts. The figure of the body which is now before us clearly shows us some parts and organs which we have not met in our study, so far. What new features do you observe on this plate?

(*Turn face section aside.*)

Harry.—There are a number of blue lines running all through the figure. Some of them are quite large and others are very small.

Fred.—There are a great many yellow and white lines running through it in all directions.

Hattie.—The head seems to be cut through the middle from the top.

T.—I am very glad for the observations which you have made. The blue lines which Harry has observed are blood-vessels of which we shall learn when we study the next chart. These " yellow and white lines " which Fred has mentioned, are parts of the great nervous system of the body. What new or strange word did you just hear me mention?

Ada.—System. You spoke of the great nervous system.

T.—By this word " system " we mean, in physiology, all those parts or organs of the body which work together for the same general purpose. We first studied the bony system. Then we studied the muscular system. Now we will study the nervous system. All these yellow and white lines represent nerves. The head, as Hattie has told us, is represented as cut through to show us the wonderful structure of the main organ of the nervous system. What do you suppose this (1) represents?

Pupils.—The brain.

T.—Right. Notice how it fills the whole of the upper part of the head. What is the upper part of the head called?

Pupils.—The skull.

T.—Yes. What other name did we give to the skull?

Pupils.—The brain-box.

T.—The brain is a very soft and delicate organ. If it were not enclosed as snugly and as safely as it is, in the skull, it would fall apart from its own weight. Really, there are two important divisions of the brain, one of which is much larger than the other, so that we sometimes speak of the large brain and the small brain. This (1) white part represents the larger brain. It fills the front and upper part of the skull. This (2) darker part represents the smaller brain. It lies be-hind and below the larger brain. When this smaller brain is cut through as represented here, its inside structure shows a beautiful figure, which, as you see (3) reminds one very much of a tree.

Frank.—Is that which passes down through the back also a part of the brain?

T.—That is called the **spinal cord.** It is composed of the same kind of substance as the brain. From the brain and from this spinal cord branch out all these nu-merous nerves which go to every part of the body. What parts of the nervous system have we now spoken of?

P.—The brain, the spinal cord and the nerves.

T.—Correct. These are the organs of the nervous system. Now I will try to help you understand some.

thing about the use of this system. What did we learn the use of the muscles to be?

P.—To move the different parts of the body.

T.—Now, it is by means of this nervous system that the mind, which largely controls the body, can tell the muscles how to act. Every muscle fiber is in connection with one of these nerve lines. At the other end each nerve is either directly connected with the brain or with the spinal cord, and through it with the brain. So you see the brain may be called the capital of the nervous system. The mind acts on the body directly through the brain. It also gets all its feelings or impressions of pain or pleasure from without, through the brain. One set of nerves runs from the brain or spinal cord to the muscles. Now, whenever a muscle is to act, every fiber of it, in some wonderful way, gets a message over its nerve line, from the brain, directing it precisely how much to contract or relax. For example, you make up your mind to close your eyes. The brain sends the mind's order over the nerve lines which go to the fibers of the circular muscle, which we have found to lie around the eye, and promptly the eye-lids close. There are two kinds of nerves. The nerves which carry the mind's messages to the muscles are called nerves of **motion.** Another set of nerves are called nerves of **feeling.** They carry the impressions from the body to the brain. These nerves are distributed so thickly near the surface of the body, in the skin, that it would be almost impossible to find a point on the body where the prick of a pin would not be felt. (*Here use the illustrations suggested on page 35.*)

But there are some movements of certain organs of

the body which the mind does not control. For instance, the beating of the heart and the breathing of the lungs must be kept up steadily as long as we live. If the muscles which produce the beating of the heart and the breathing movements of the chest must always be directed by the mind, we could think of nothing else, and it would be death to fall asleep. So the Creator has wisely provided that certain parts of the nervous system constantly and faithfully direct all such muscular operations which must be kept up constantly, without any attention from the mind.

EXERCISE XII.

MANIKIN OF THE HEAD.

T.—I am now going to show you a very interesting illustration of the location and appearance of the brain. Here we have a collection of five manikins. A manikin is a skillful arrangement of the parts or organs of the body in their proper natural order and places, so that we are enabled by the use of such a manikin to take the body apart, as it were, to see how each part is located and how it appears. This is a manikin of the head. What does this first outside figure show?

P.—The face and hair.

T.—Now we will take off the skin and hair by laying back this section. (*Lay section back.*) How different the head looks when the skin and hair are taken off! Addie, will you tell us what you notice about this view of the head?

Addie.—It shows a good many muscles.

T.—So it does. Some of these muscles of the head lie

very clo-: under the skin, while some are more deeply located These blue and bright-red parts represent blood-v ssels, of which we shall learn later on. Now we will r move these muscles and blood-vessels by laying this section aside. (*Turn section.*) What do you observe now?

Charles.—I think those must be the bones of the skull.

T.—You are right, Charles. What else does this section show?

P.—Some blood-vessels and nerves.

T.—What did we start out to find?

P.—The brain.

T.—Where did we learn the brain to be located?

P.—In the skull.

T.—So let us now remove these skull bones. *Lay section aside).* Here the brain appears; but the large brain is here shown as cut in two, while the smaller brain (u) shows its natural outside appearance. Let us get a deeper view of it. (*Turn section*) Here we see the brain as it appears when cut through the middle of the head from front to back. Notice how wavy its surface appears, how the blood-vessels run all through it, and how much like the figure of a tree the surface of the small brain appears, when cut through. What do you suppose this (124, 150) large cord of nervous matter is, which connects with the brain above and passes down through the back bone?

P.—The spinal cord.

T.—Now let us carefully notice one more thing before we turn from this manikin. See how the nerves pass out from the brain to go to various parts and organs—some to the eye, some to the nose, some to the tongue and teeth, and others to other parts. I wish you especially to remember this last observation, as it will help you the better to understand our next lesson.

EXERCISE XIII.

T.—We have learned that the nerves of feeling are found in all parts of the body. So we say the sense of feeling is distributed over the whole body. But we have four other kinds of nervous sensation besides ordinary feeling. Altogether, they are called the five senses. They are **feeling, hearing, seeing, tasting,** and **smelling.** All of these except feeling are called special senses, because each has a special organ and a special nerve to receive a special impression and carry it to the brain to be perceived by the mind. I will first try to teach you something about the sense of hearing.

(*Referring to the diagram of the ear on chart of nervous system.*) What does this (1) seem to represent?

P.—An ear.

T.—Yes; this is a view of the whole structure of the ear. It shows not only this (1) outside part of the ear, which we are so accustomed to see, but also this part (2) called the ear-tube, which leads into the temporal bone of the head, in which these (3–17) wonderful parts of the inner ear are so safely hidden away from danger. What was the last thing I asked you to observe when we studied the brain, in the manikin of the head?

P.—That nerves go directly out from the brain to different parts.

T.—Right. Those nerves which we have spoken of as nerves of special sensation all go out from the brain. One of these is the special nerve of hearing. It goes from the brain to the ear. Here (*yellow-colored nerve branch*) we see a part of this nerve of hearing connecting with the apparatus of the ear. This nerve is affected only by **sound**. But in order that it may be properly impressed by sounds, a special instrument is necessary to carry the sound in a proper manner to the nerve. This instrument is the ear.

Charles—A part of the ear reminds me very much of a funnel.

T.—Yes; we may call it a sound funnel. But the most wonderful structure of the ear is this (3–17) inside apparatus. When you are somewhat farther advanced in physiology, you will be prepared to understand the structure and use of all these delicate parts of the ear. For the present, please remember only this: that the sounds which are collected by the ear-funnel and tube are passed through these tubes and chambers of the inner ear in such a way that a proper impression is made on this nerve of hearing. This nerve then conveys such impressions to the brain, where they are received by the mind. It is in this way that the mind receives through this nerve and its instrument, the ear, all the delights of music, the instruction of spoken words, the songs of birds, and the roar of the thunder, with all the hundreds of various sounds and noises which we hear every day.

Now, Charles, will you tell us what you understand to be the use of the ear?

Charles.—To gather sound-waves and carry them to the nerve of hearing.

T.—Correct.   Fannie, what is the use of the nerve of hearing?

Fannie.—To carry the impression of sound to the brain.

T.—Is the nerve of hearing affected by anything else besides sound-waves?

P.—It is not.

T.—So we call it what kind of a nerve?

P.—A nerve of special sensation.

T.—Now we will talk about another very important nerve of special sensation.   This figure (*Eye figure. Chart of Nervous System*) represents the eye cut through so as to give us a view of its inside parts. The eye is the instrument of sight.   As the nerve of hearing passes from the brain to the ear, so the nerve of sight passes from the brain to the eye.   What must come to the eye in order that we may see?

P.—Light.

T.—Yes; the nerve of sight is not at all affected by sound, and the nerve of hearing is not affected by light.   We found how carefully the ear is hidden away from danger in the little caves of the temporal bone of the head.   So, by looking each other in the face, we observe how carefully the eye is lodged in hollow places among the upper bones of the face and under the over-hanging bone of the forehead.   But what else do you see over the eye for its protection?

P.—The eyebrows.

T.—And what more?

P.—The eyelids.

T.—Anything more?

P.—The eyelashes.

T.—Yes; all these are for the protection of the eye.
(Explain how. See "Protection of the Eye," page
42.)

There are a number of muscles attached to the eyeball. By the action of these muscles, the eye is properly directed to the objects which we wish to see. Then
a part of the light which falls upon the eye enters
through the little opening which you see in the center
of the eyeball when you look directly into another person's eye. Then the wonderful parts of the inside of
the eye affect the light in such a way as to make a picture of the thing or things we see, on the br ad end of
the nerve of sight on the inside of the back part of the
eye. This tiny picture which is thus made, so affects
the nerve that it carries an impression to the brain from
which the mind gets a correct view of the appearance
of the object from which the light came to the eye.
When you are farther advanced, you will be able to
understand all the parts of the eye, by which this wonderful picture-making is done.

(If the teacher thinks it advisable, he can refer to the
manikin of the eye, and use the description found on
pages 42-47.)

Another special nerve is the nerve of taste. It goes
from the brain to the tongue and other parts of the
mouth. The tongue is the chief instrument of taste.
This figure (*Figure of " Sense of Taste," Nervous Chart*) shows how the very numerous branches
of the nerve of taste go to all parts of the tongue and
mouth. These nerves are affected by the taste or flavor
of things which are brought into the mouth, or touched
with the tongue.

Now we have spoken of three of the four special nerves. Can you name them?

P.—The nerve of hearing, the nerve of sight, and the nerve of taste.

T.—What do you suppose is the **fourth** special nerve?

P.—The nerve of smell.

T.—Yes; and I feel sure that you can tell me what the special instrument of this nerve is?

P.—The nose.

T.—Correct. This figure (" *Sense of Smell*") represents the nose cut in two, to show us what a large number of branches of the nerve of smell are spread all over its inside parts.

(Review this exercise well.)

EXERCISE XIV.

The teacher will find it to be an excellent plan to make this exercise a black-board outline review. Such an outline review impresses the main facts learned strongly upon the pupils' minds. By careful questioning, draw out from the pupils the leading facts learned about each of the three great systems which have so far been studied, and arrange them on the board, in outline form, as near as may be, as follows:

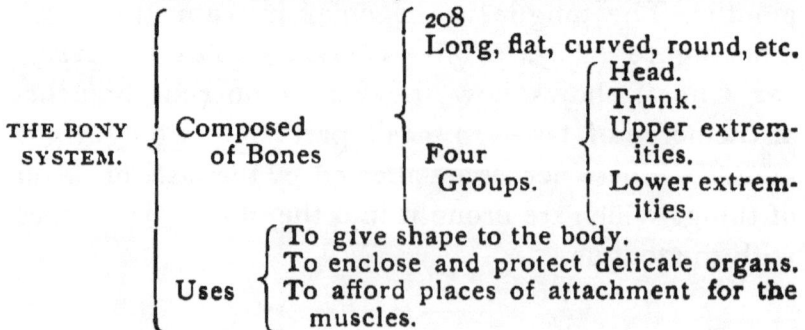

THE BONY SYSTEM.	Composed of Bones	208 Long, flat, curved, round, etc.	
		Four Groups.	Head.
			Trunk.
			Upper extremities.
			Lower extremities.
	Uses	To give shape to the body.	
		To enclose and protect delicate organs.	
		To afford places of attachment for the muscles.	

THE MUSCULAR SYSTEM.
{
Composed of Muscles.
{
Lean meat.
Have many fibers.
Power of contraction.
Attached to the bones.
}

Use { To move the different parts of the body.
}

THE NERVOUS SYSTEM.
{
Composed of
{
Brain.
Spinal cord.
Nerves.
}

Uses
{
To serve the mind and brain in the control of the action of the muscles.
To give us sensations of touch or feeling, sound, sight, taste and smell.
}
}

EXERCISE XV.

CHART OF VEINS AND ARTERIES.

T.—How many and what great systems of the body have we so far studied?

P.—Three; the bony system, the muscular system and the nervous system.

T.—What new features do you see in the figure of the body before us?

P.—There are a great many blue and red tubes running in all directions through it.

T.—We merely noticed some such blue lines on the chart which we studied before. What did we call them?

P.—Blood-vessels.

T.—This chart shows us quite a number of the blood-vessels of the body. The blue vessels represent one kind of blood-vessels and the red another kind. Why should they be called blood-vessels?

Mary.—Because they contain blood.

T.—Right; the blood in our bodies is constantly

flowing through these blood-vessels, to and from every part and organ. Let me try to explain to you why the blood is constantly flowing through the body in these blood-vessels. (*Here use, as an explanation, the first two paragraphs of the circulatory system, pages 55–56.*)

The chief organ of the blood-circulating system is the **heart** (S). Notice where the heart is situated in the body. What do you observe about its location?

George.—It is in the upper part of the trunk.

T.—Can some one name its place more precisely than that?

Fannie.—It is in the chest.

T.—Harry, can you describe its position still more accurately?

Harry.—I should say that the heart lies a little below and to the left of the middle of the chest.

T.—Very good. Now since you have made such a fine observation, I will show you at another place, exactly how the heart is placed in the body. (*Turn to body manikin and show its position.*)

Ralph—How large is a person's heart?

T.—A man's heart is about as large as his fist. It is a very strong muscular organ. It has four hollow places in it. These are called chambers or blood pockets. Two of these chambers are on the right side of the heart and two are on the left side.

The use of the strong muscle walls of the heart is to drive the blood in its course to all parts of the body from the top of the head to the tips of the toes. The blood from the right side of the heart is driven through blood-vessels to the lungs, to be purified, as we shall soon learn more particularly. From the left side of the heart, the blood is carried through blood-vessels to all parts of the body to carry nourishment—that is, building and repairing material—wherever it is needed.

These **red** blood-vessels, large and small, (15, 71), are **arteries.** These arteries carry **pure** blood from the heart to all parts of the body. This great artery (15) which starts from the heart is called the **aorta.** It branches off like a tree into many smaller arteries. What do you suppose to be the use of these arteries (19, 21) which pass up through the neck?

P.—To carry pure blood to the head.

T.—Correct. Please notice into what a wonderful net-work of fine arteries these neck arteries divide all through the head. What seems to be the use of this artery (55)?

P.—To carry pure blood to the arm.

T.—Right; notice how it branches off into many smaller branches. A little above the heart—about here (17)—the great aorta artery bends downward through the inside of the body, as we see it here (34). Then here in the lower part of the abdomen it divides into two great branches, for what purpose, do you suppose?

P.—To send arteries into both of the lower limbs.

T.—Yes; you now have some idea how the pure blood gets from the heart to all parts of the body through the arteries.

As the blood passes through the body it becomes impure, as you will understand more fully hereafter. This impure blood is gathered up by the veins and carried back to the heart. The heart then drives it on to the lungs where it is purified.

purities in exchange. Then it is gathered up by the veins. In the lower part of the body these veins gather into this (51) large ascending vein. In the upper part of the body they gather into this (27) large descending vein. Finally, both these large veins empty into the right upper chamber of the heart.

(*Drill the pupils thoroughly on this full route of the blood until they can individually describe it.*)

EXERCISE XVII.

CHART OF VEINS AND ARTERIES.

T.—In our last lesson we learned about the course of the blood through the body. What change comes over it as it passes through the system?

P.—It becomes impure.

T.—What vessels gather it up after it has become impure?

P.—The veins.

T.—Where do the veins carry it to?

P.—To the right upper chamber of the heart.

T.—What seems necessary to be done before it goes through the body again?

P.—It must be purified.

T.—Yes; and this purifying of the blood is performed in the lungs. What carries the impure blood from the heart to the lungs?

P.—A large artery.

T.—Yes; this (16) is the artery you speak of. We see here how the lungs are situated (R), and how the artery which brings the impure blood from the heart, sends its branches all through them. What can you say of the location of the lungs?

Edward.—They are in the chest.

Nellie.—They lie on both sides of the heart.

T.—Yes, or perhaps better said, over and around the heart. These two organs—the lungs and the heart—fill the whole cavity of the chest. Let us now turn to the manikin of the body and get a still more interesting view of the place and surroundings of the lungs.

(Turn to the body-manikin. As you lay aside the outer parts slowly, speak of removing the skin, then the outer muscles, then the ribs, when the lungs will lie before you. Then, while the lungs are thus exposed before the class, speak of their structure, and of the wind-pipe and air-tubes as given on pages 70–71.)

Here, in the lungs, the oxygen of the air which we breathe unites with the blood. At the same time the impurities of the blood pass from it into these air tubes, and are thrown off with the breath which escapes. In this way the blood is rid of the impure matter, and changed from its dark color to a bright red, and then is again fit to be driven by the heart to all parts of the body. Do you remember how the blood gets back to the heart from the lungs?

P.—It is carried back by large veins.

T.—To which side of the heart?

P.—To the left side.

T.—Through what vessels does the heart then drive it to all parts of the body?

P.—Through the arteries.

EXERCISE XVIII.

It is suggested that this exercise be devoted to a familiar conversational lesson on **proper breathing** and

good ventilation. Explain how we breathe as taught on page 73. Show that anything which interferes to prevent taking in full breaths of air is injurious. (See " *Health of the Respiratory Organs,*" page 74.) In the lungs the life-giving oxygen is taken away from the air. To the unused portion of the air are added the impurities discharged from the blood in the lungs. Show the consequence of this foul, cast-off breath accumulating in a room without sufficient supply of pure, fresh air. Impress thoroughly and illustrate well, from your own knowledge, and in your own way, the important principles of good ventilation. (*See also* " *Impurities of the Breath,*" *and* " *Ventilation* " *on page 76*).

EXERCISE XIX.

T.—While we were learning about the blood— its circulation through the body, and its purification in the lungs—this question must have come to your minds: Where does the body-building and body-repairing material which the blood carries to all parts come from? Can any one of you give me the proper answer to this question?

Harry.—It comes from the food we eat.

T.—You are quite right, Harry. But blood is so different from the food which we eat, that great changes must be necessary before the material of our food is fit to be laid up in our bodies as muscle, or bone, or nervous matter. Besides, not all of the food which we eat is fit for body-building or body repairing. Some parts of it are quite useless. The separation of such useless food-portions from the useful parts, and the proper preparation of the useful parts for body material is called **digestion**. Digestion is one of the most important processes which take place in the body. Upon a good digestion of our food, much of our com. fort, our health and our very lives depend. So the Creator has furnished the body with a large number of important organs to perform this work, step by step. The organs which perform the work of digestion are called the digestive system. Now let us see what we can learn about the manner in which the food is digested.

(Refer to Chart of the Circulation.)

T.—Where is the food placed in eating?

P.—Into the mouth.

T.—Yes; in the mouth the first and second steps of digestion are performed. What happens to the food in the mouth?

P.—It is chewed by the teeth.

T.—Right. This is the first step of digestion. But while the teeth are chewing the food, it is mixed with saliva. This is a watery-like juice which flows into the mouth from little organs which lie at different places near the mouth. These little organs prepare this saliva from the blood for this very purpose of assisting in digestion—that is, in preparing the food for the nourishment of the body. This mixing of the food with saliva is the second step in digestion. Now when the food has been properly chewed and mixed with saliva, in the mouth, what do you suppose is the third step?

George.—I should think it would then be ready to be swallowed?

T.—You are right, George. In swallowing the food, it passes down through this (1) tube into this organ (2). Do you know the name of this organ?

P.—The stomach.

T.—This gives us a good view of the shape of the stomach and also of its outside and inside appearance. But we will turn to the body-manikin to see exactly where the stomach lies. (*Do so.*)

T.—In the stomach the food is much changed. How this change is produced, you will learn more fully, later on. The change which takes place in the food in the stomach is called the fourth step of digestion.

By the time this fourth step of digestion is reached, some of the more watery portions of the food are ready to be taken into the blood. So numerous little veins (*See blue veinlets on stomach figure*) which are distributed throughout the stomach walls, take up these prepared food portions and carry it directly into the circulatory system.

The remaining undigested food parts now pass on to the fifth step of digestion. They pass out through this (5) end of the stomach through a curious little muscle gate which refuses the food to pass out until it is ready for this new step in the process. As it passes out of the stomach, the food enters the intestines (4). Here it is still more changed by being acted upon by two substances which are also specially prepared from the blood, to help in this work of digestion. One of these two substances comes from this organ (*turn to body-manikin*) lying back of the stomach, called the pancreas (59). The other substance, called bile, is made from the blood by the liver. Here lies this large organ toward the right side of the body over the stomach. Notice the large blood-vessels and the bile-sac within the liver.

Here, in the upper part of the intestines, the useful part of the food is finally separated from the useless part and passes out of the intestine walls through little tubes and veins which carry it to the circulatory system, which distributes it to all parts of the body, where needed, as we have learned before.

EXERCISE XX.

Let this exercise be a blackboard outline review on the last three systems learned. Draw out the facts which, when written on the board, will stand as follows :

THE CIRCULATORY SYSTEM

Composed of
- Heart.
- Arteries.
- Veins.
- Capillaries.

Use
- To carry the blood to all parts of the body.

THE BLOOD PURIFYING SYSTEM.

Composed of
- Lungs.
- Wind-pipe.
- Air-tubes.

Use
- To purify the blood.

THE DIGESTIVE SYSTEM.

Composed of
- Mouth.
- Teeth.
- Tongue.
- Salivary glands.
- Stomach.
- Liver.
- Intestines.
- Pancreas.
- Other organs.

Use
- To prepare the food for the nourishment of the body.

ADDITIONAL SUGGESTIONS.

Having now suggested quite a number of exercises indicative of a practical method of conducting elementary oral drills on six of the great systems of the body, it is believed by the authors, that by the time these exercises have been used, the teacher will have so well acquired the "run" of the method, that it will be unnecessary to add any farther exercises in this book. The teacher can draw plenty of facts from the lessons in the main part of this book, from the Aid and from other sources, for the illustration of any subject which is chosen for any exercise. However, for the purpose of aiding the teacher in making such a choice of lesson-subjects as will be appropriate and in proper order, the following additional exercises are suggested:

Exercise xxi. A familiar conversational talk on the health of the digestive organs; on eating too fast, too much or too frequently; on eating indigestible food, and on the favorable effects of gentle exercise on digestion.

Exercise xxii. On the structure of the skin—its two layers, blood-vessels, numerous nerve endings, perspiration glands and tubes, fat-cells, oil-glands, coloring-matter and hair growth, as illustrated in the upper left-hand figure on the chart of the circulation, and described in this book on pages 112–117.

Exercise xxiii. On the perspiration of the skin, by which a great amount of worn-out and poisonous matter is expelled from the body, and on the need of clean, clothes and frequent bathing—that is, bodily cleanliness—to avoid disease and death. (*See pages 117–120.*)

EXERCISE XXIV. On the meaning of strong drink; how the different kinds of wine are made, and that they all are made "strong" by the alcohol which they contain. (*See pages 122–126.*)

EXERCISE XXV. How the beers are made and that the harmful substance in each kind of beer is alcohol. (*See pages 127-128.*)

EXERCISE XXVI. How the stronger liquors are produced, and wherein they differ from the wines and beers. (*See pages 128–129.*)

EXERCISE XXVII. On the nature of alcohol—its physical properties, inflammability, low freezing point, greed for water and its consequent tendency to harden and destroy many parts of the body. (*See pages 123–124.*)

EXERCISE XXVIII. Effects of alcohol upon the body. First effect of physical excitement and second effect of physical depression. (*See pages 130–134.*)

EXERCISE XXIX. The drunken stage—the cause of the drunkard's stammering, staggering and squinting. (*See pages 134–135.*)

EXERCISE XXX. The "dead-drunk" condition. (*See pages 135–136.*)

EXERCISE XXXI. Effects of alcohol on the stomach. (*Refer to stomach plates and see pages 142–145.*)

EXERCISE XXXII. Effects upon the brain. (*Refer to intemperance plates and see page 136.*)

EXERCISE XXXIII. Effects upon the liver and kidneys. (*See figures of these organs on intemperance plates and pages 145–146.*)

EXERCISE XXXIV. Tobacco and its effects upon the body. (*See intemperance plates and pages 147–153.*)

HOW TO USE THE AID.

For the benefit of those teachers who desire to use the Teachers' Anatomical Aid to the very best advantage to themselves and their pupils, we offer the following suggestions:

1. For the purpose of an elementary oral drill of a class of beginners, the Aid will enable you to conduct the exercise on the "object lesson" plan. By this method, such class drills are made interesting and instructive, and the knowledge which the pupils thus acquire, though it is simple and elementary, will prove to be a well-laid stepping-stone to their later text-book study of physiology.

2. During such a course of elementary oral instruction, the pupils use no book for study. They should come to the class-drill to learn the facts of the lesson which is taught, *from the plain, practical statements of the teacher, and from their own observation of the objects brought before them on the charts of the Aid, for the illustration of such facts and statements.*

3. This Manual will suggest to you, step by step, the proper order of such elementary lesson. You can safely follow the course of exercises for which it furnishes you material and chart references. However limited your experience may be, the material for an oral lesson, from day to day, is herein placed in your possession. With the book in your hands, and the Aid before you, you will be able to make a thorough private preparation, to fill your own mind with the facts pertaining to the subject of the intended exercise, and then, filled with enthusiasm, as well as information, to go before your pupils and present the lesson with confidence and success, *without handling a book during such class exercise.*

4. Do not attempt too much at any one exercise. Adapt both the quality and quantity of your instruction to the age and capacity of your pupils. A few facts, taken in their proper order, and well taught, are worth more than many facts poorly taught by the teacher, and consequently, poorly understood and soon quite forgotten by the pupil.

5. When this Manual is used as a text-book by the pupils, the teacher should assign for each lesson only so many topics as the pupils can thoroughly master. It is suggested that at recitation each subject should first be described, "topically," as accurately and as fully as possible, with such illustrative references to the Aid, by the pupil, as the subject discussed calls for. This should be followed by such questions from the teacher as will bring out any omitted facts. Then, in farther practical illustration of the facts and principles stated, or for the enforcement of any hygienic precepts pointed out in the lesson, the teacher will add such information as he may have at command, and which in his judgment may be appropriate and valuable. Finally, for the purpose of "fixing" in the minds of all, and arranging in proper order, the points of the topic in hand, let the teacher, by a series of review questions, *draw out* what has been learned, in answers from the whole class.

6. We will here repeat a suggestion which appears in the preface of this book, namely: that in the preparation of the lessons by the pupils when using this Manual as a text book, they must have access to the Aid. While such access to the Aid, by the pupils, is absolutely necessary in preparing a lesson from this book, it is also very desirable in studying a lesson from any other text book. By a little wise planning the teacher can always provide such opportunity for reference to the proper charts on the part of the members of the physiology class, without any disturbance of other pupils.

7. During a drill exercise. or recitation, the Aid should be brought close before the pupils, at a convenient elevation and under proper conditions of light, so that all may clearly see any part or organ to which reference is made by way of illustration.

INDEX

COMPLETE NERVOUS SYSTEM

(Referring to Fifth Chart of Aid.)

I. THE CRANIAL AND SPINAL SYSTEM.

No.	Common Name.	Latin or Professional Name.
1	Brain.	Cerebrum.
2	Small Brain.	Cerebellum.
3	Tree of Life.	Arbor vitæ.
4	Varol's Bridge.	Pons Varolii.
5	Three fold Nerve.	Nervus Trigeminus.
6	Abducent Nerve.	Nervus abducens.
7	Face and Sound Nerve.	Nervus facialis et acousticus.
8	Tongue and Pharynx Nerve.	Nervus glosso-pharyngeus, vagus et accessorius.
9	Willis' Accessory Nerve.	Nervus accessorius Willisii.
10	Loose cavity containing lung and stomach nerve ducts.	Nervus vagus pneumo-gastricus.
11	Lower tongue nerve.	Nervus hypoglossus.
12	Descending branches hypoglossal nerves.	Rami descendens nervi hypoglossi.
13	Cranial portion of the Spinal Cord.	Medulla oblongata.
14	Decussate pyramid.	Decussatio pyramidum.
15	Part of cervical spinal cord.	Pars cervicalis medullæ spinalis.

No.	Common Name.	Latin or Professional Name.
16	Part of thoracic spinal cord.	Pars thoracica medullæ spinalis.
17	Bulbous expansion at end of spinal cord.	
18	Terminal threads, spinal cord.	Filum terminale.
19	Neck nerves 1.	Nervus cervicalis 1.
20	Neck nerves 8.	Nervus cervicalis 8.
21	Network of neck nerves.	Plexus cervicalis.
22	Network of arm nerves.	Plexus brachialis.
23	Back bone nerve 1.	Nervus dorsalis 1.
24	Back bone nerve 2.	Nervus dorsalis 2.
25	Nerves between ribs.	Nervi intercostales.
26	Loin nerve 1.	Nervus lumbalis 1.
27	Loin nerve 5.	Nervus lumbalis 5.
28	Network of loin nerves.	Plexus lumbalis.
29	Anterior crural nerve.	Nervus cruralis anterior.
30	Hip—abdominal nerve.	Nervus ilio — hypogas tricus (ramus exterior et interior).
31	Hip—groin nerve.	Nervus ilio inguinalis.
32	Groin skin nerve.	Nervus inguino cutaneous.
33	Obturator nerve.	Nervus obturatorius.
34	Sacral nerve 1.	Nervus sacralis 1.
35	Sacral nerve 5.	Nervus sacralis 5.
36	Network of sacral nerves.	Plexus sacralis.
37	Coccyx nerves.	Nervi coccygei.
38	Sympathetic nerve.	Nervus sympaticus.
39	Upper cervical ganglion.	Ganglion cervicale superior.
40	Middle cervical ganglion.	Ganglion cervicale medium.
41	Lower cervical ganglion.	Ganglion cervicale inferior.
42	Thoracic ganglia.	Ganglia thoracica.
43	Loin ganglia.	Ganglia lumbalia.
44	Sacral ganglia.	Ganglia sacralia.
45	Coccyx ganglion.	Ganglion coccygeum.
46	Connecting branches between sacral and sympathetic nerves.	

No.	Common Name.	Latin or Professional Name.
47	*Sciatic nerve* (Hip nerve).	Nervus ischiadicus.
48	Groin nerve.	Nervus inguinalis.
49	Shoulder bone nerves.	Nervi supraclaviculares.
50	Branches of skin and axle nerves.	Rami cutaneus et nervi axillaris.
51	Arm skin nerves, internal posterior.	Nervi cutane us brachii, internus posterior.
52	Arm skin nerve, small internal.	Nervi cutaneus brachii internus (minor).
53	Branches middle skin nerves (arm).	Ramus nervi cutanei medii.
54	Middle arm skin nerve.	Nervus cutaneus brachii medius v. internus major.
55	Branches skin palm nerves of middle skin nerve.	Ramus cutaneus palmaris, nerv. cutan. medii.
56	Branches of middle under skin nerve over ulna.	Ramus cutaneus ulnaris nerv. cutan. medii.
57	Branches under skin nerves, overlying cutaneous muscle.	Ramus cutaneous nerv. musculo cutanei.
58	Radial nerve branches.	Ramus nervi radiales.
59	Voluntary ulna nerve.	Nervus ulnaris volaris.
60	Voluntary finger nerves.	Nervus digitales volaris.
61	Network of arm pit nerves.	Plexus axillaris (brachialis).
62	Middle nerve, sending branches to thumb, index and middle finger and radial side of ring finger.	Nervus medianus.
63	Ulna nerve.	Nervus ulnaris.
64	Voluntary ulna nerve,	Nervus ulnaris volaris.
65	Spiral muscular nerve lying against radius.	Nervus musculo—spiralus v. radialis.
66	External elbow joint nerve.	Nervus interosseus externus.
67	Superficial radial nerve.	Nervus radialis superficialis.
68	Musculo cutaneus nerve.	Nervus musculo—cutaneus.
69	Anterior leg nerve.	Nervus cruralis anterior.
70	External anterior femoral nerve.	Nervus cutaneus femoris anterior externa.

No.	Common Name.	Latin or Professional Name.
71	Groin nerve.	Nervus inguinalis.
72	Groin skin nerve.	Nervus inguino cutaneus.
73	Large saphenic nerve.	Nervus saphenus major.
74	Middle anterior femoral nerve.	Nervus cutaneus femoris anterior medius.
75	Internal anterior femoral nerve lying against small saphenic.	Nervus cutaneus femoris anterior internus v. saphenus minor.
76	Branches of hip abdominal nerves.	Rami nervi ilio hipogastrici.
77	Branches hip groin nerves.	Rami nervi ilio inguinalis.
78	Branches muscular leg nerves.	Rami musculares nervi cruralis.
79	Superficial peroneal nerve fibular.	Nervus peronæus superficialis, con.
80	Internal foot skin nerve.	Nervus cutaneus dorsi pedis internus, et.
81	Middle foot skin nerve.	Nervus cutaneus dorsi pedis medius.
82	External leg skin nerve.	Nervus cutaneus cruris externus.
83	Deep peroneal or fibular nerve.	Nervus peronæus profundus.
84	Deep branch of peroneal nerve.	Ramus internus peronæus profundus.
85	External branch of peroneal nerve.	Ramus externus peronæus profundus.
86	Cervical or neck back bone joint.	Vertebra cervici (7).
87	Back bone joint (1).	Vertebra dorsi (1).
88	Back bone joint (12).	Vertebra dorsi (12).
89	Loin back bone joint (1).	Vertebra lumbalis (1).
90	Loin back bone joint (2).	Vertebra lumbalis (2).
91	Sacrum bone.	Os sacrum.
92	Coccyx bone.	Os coccygi.
93	First rib.	Costa prima.
94	Last rib.	Costa termina.
95	Crest of ilium bone.	Crista ossis ilii.

No.	Common Name.	Latin or Professional Name.
96	Underlying collar muscle connecting with sternum.	Musculus sterno — cleido mastoidens.
97	Front scalene muscle.	Musculus scalenus anticus.
98	Middle scalene muscle.	Musculus scalenus medins.
99	Internal intercostal muscles.	Musculi intercostales interni.
100	External intercostal muscles.	Musculi intercostales externi.
101	Square loin muscle.	Musculus quadratus lumborum.
102	Large loin muscle.	Musculus psoas major.
103	Internal iliacal muscle.	Musculus iliacus internus.
104	Deltoid muscle (shoulder).	Musculus deltoideus.
105	Large breast muscle.	Musculus pectoralis major.
106	Flexible forearm muscle.	Musculus biceps flexon cubiti.
107	Fold in forearm.	Plica cubiti.
108	Head of ulna.	Caput ulnæ.
109	Aponeuroses of the palm.	Aponeurosis palmaris.
110	Fleshy ball of thumb.	
111	Short palm muscle.	Musculus palmaris brevis.
112	Cephalic vein (arm).	Vena cephalica brachii.
113	Basilical vein.	Vena basilica.
114	Middle basilical vein.	Vena mediana basilica.
115	Middle cephalic vein.	Vena mediana cephalica.
116	Head of humeris bone.	Caput ossis humeri.
117	Sharp process of scapula.	Processus coracoideus.
118	Deltoid muscle.	Musculus deltoideus.
119	See No. 105.	
120	Small breast muscle.	Musculus pectoralis minor.
121	Flexible muscle of biceps.	Musculus biceps flexor cubiti.
122	Short head of biceps muscle.	Caput breve, musculus bicipitis.
123	Long head of biceps muscle.	Caput longum, musculus bicipitis.
124	Coracoid arm muscle.	Musculus coraco-brachialis.
125	Internal arm muscle.	Musculus interna brachialis.

No.	Common Name.	Latin or Professional Name.
126	Internal head of extending triceps muscle.	Caput internum, m. tricipitis extensoris.
127	Long head of extending triceps musole.	Caput longum m. tricipitis extensoris.
128	Muscle, serving to turn palm of hand upwards.	Musculus supinator longus.
129	Muscle, long. round extending wrist.	Musculus extensor carpi radialis longus.
130	Muscle, serving to turn palm of hand downward.	Musculus pronator teres.
131	Round wrist muscle bending or turning.	Musculus flexor carpi radialis.
132	Short, like functions as 128.	Musculus supinator brevis.
133	Common bending finger muscles.	Musculi flexores, digitorum communes.
134	One of the wrist bending muscles.	Musculus flexor carpi ulnaris.
135	Long bending striking muscle.	Musculus flexor pollicis longus.
136	Muscles serving thumb.	Musculus abductor et flexor brevis pollicis.
137	Drawing thumb to the index finger.	Musculus abductor pollicis.
138	Shoulder artery.	Arteria axillaris.
139	Arm arteries and veins.	Arteriæ et venæ brachialis.
140 141	} Arteries and veins of ulna.	Arteriæ et venæ ulnaris.
142	Upper anterior spine of ilium.	Spina ilii anterior superior.
143	Tailor's muscle.	Musculus sartorius.
144	Middle gluteal muscle (serving to turn thigh in and outward).	Musculus glutæus medius.
145	Deep leg stretching muscle.	Musculus tensor faciæ latæ.
146	Straight femoral muscle.	Musculus rectus femoris.
147	External vastus muscle.	Musculus vastus externus
148	Muscle, serving to bring thigh together.	Musculus pectinæus.

No.	Common Name.	Latin or Professional Name.
149	Long drawing muscle.	Musculus abductor longus.
150	Large drawing muscle.	Musculus abductor magnus.
151	Leg muscle.	Musculus cruralis.
152	Internal vastus muscle.	Musculus vastus internus.
153	Tendon extending leg.	Tendo extensorius cruris.
154	Knee.	Patella.
155	Shin.	Tibia.
156	Internal ⎱ Ankle joint pro-	Malleolus internus.
157	External ⎰ jections.	Malleolus externus.
158	Transverse ligament.	Ligamentum transversum.
159	Foremost tibial muscle.	Musculus tibialis anticus.
160	Muscle, extending toes and foot.	Musculus extensor digitorum pedis longus.
161	Long peroneal muscle (Fibula).	Musculus peronæus longus.
162	Short peroneal muscle.	Musculus peronæus brevis.
163	Long extending striking foot muscle.	Musculus extensor pollicis pedis longus.
164	Counteracting on 160.	Musculus extensor digitorum pedis brevis.
165	Short striking foot muscle.	Musculus extensor pollicis pedis brevis.
166	Sole muscle.	Musculus soleus.
167	Femoral artery.	Arteria femoralis.
168	Femoral vein.	Vena femoralis.
169	Large saphenic vein.	Vena saphena magna.

II. THE SYMPATHETIC SYSTEM.

DISTRIBUTION OF FACIAL AND PNEUMOGASTRIC NERVES.

1	Descending thoracic aorta.	Aorta descendens thoracica.
2	Innominate artery.	Arteria innominata.
3	Right under-shoulder artery.	Arteria subclavia dextra.

No.	Common Name.	Latin or Professional Name.
4	Right carotid artery.	Arteria carotis communis dextra.
5	Internal carotid artery.	Arteria carotis interna.
6	External carotid artery.	Arteria carotis externa.
7	Upper thyroid artery.	Arteria thyroidea sup.
8	External jaw artery.	Arteria maxillaris externa (v. facialis).
9	Occipital artery.	Arteria occipitalis.
10	Upper ear artery.	Arteria auricularis superior.
11	Temporal artery.	Arteria temporalis.
12	Pulmonary arteries and veins.	Arteriæ et venæ pulmonales.
13	Intercostal arteries and veins.	Arteriæ et venæ intercostalis
14	Descending aorta (abdominal) with lower aortic plexus.	Aorta descendens abdominalis, con plexus aorticus inferior.
15	Cœliac artery and plexus.	Arteria cœliaca con plexus cœliacus.
16	Kidney artery and plexus.	Arteria renalis con plexus renalis.
17	Upper mesenteric artery with plexus.	Arteria mesenterica superior con plexus mesentericus sup.
18	Lower mesenteric artery with plexus.	Arteria mesenterica inferior con plexus mesentericus inferior.
19	Common iliacal artery.	Arteria iliaca communis.
20	Network of upper abdominal nerves.	Plexus hypogastricus superior.
21	Network of hæmorrhoidal nerves.	Plexus hæmorrhoidales.
22	Network of nerves surrounding bladder.	Plexus vesicalis.
23	Network of prostate nerves.	Plexus prostaticus.
24	Network of lower abdominal nerves.	Plexus hypogastricus inferior.
25	Lower phrenic arteries with network of phrenic nerves.	Arteriæ phrenicæ inferiores con plexus phrenicus.

No.	Common Name.	Latin or Professional Name.
26	Great network of stomach nerves.	Plexus gastricus magnus.
27	Splenic artery with network of splenic nerves.	Arteria splenica con plexus splenicus.
28	Liver artery with network of liver nerves.	Arteria hepatica con plexus hepaticus.
29	Upper network with semi-lunar ganglion.	Plexus solaris con ganglion semi-lunarius.
30	Loin ganglion.	Ganglion lum bale.
31	Sacral ganglion.	Ganglion sacrale.
32	Thoracic gland 1.	Ganglion thoracicum 1.
33	Thoracic gland 7.	Ganglion thoracicum 7.
34	Large splanchnic nerve.	Nervus splanchnicus major.
35	Small splanchnic nerve.	Nervus splanchnicus minor.
36	Upper network of thoracic nerves.	Plexus thoracicus superior.
37	Lower ganglion of neck nerves.	Ganglion cervicale inferior.
38	Middle ganglion of neck nerves.	Ganglion cervicale medina.
39	Upper ganglion of neck nerves.	Ganglion cervicale superior.
40	Network of nerve molles.	Plexus nervorum mollium.
41	Front ear nerve.	Nervus auricularis anterior.
42	Posterior ear nerve.	Nervus auricularis posterior.
43	Facial nerves and branches causing goose flesh on skin.	Nervus facialis et pes anserius.
44	Small occipital and upper ear nerve.	Nervus occipitalis minor et nervus auricularis superior.
45	Willis' accessory nerve.	Nervus accessorius Willisii.
46	Network of neck nerves.	Plexus cervicales.
47	Vagus nerve.	Nervus vagus.
48	Recurrent nerve.	Nervus recurrens.
49	Phrenic nerve.	Nervus phrenicus.
50	Network of arm nerves.	Plexus brachialis.
51	Network of loin nerves.	Plexus lumbalis.
52	Network of sacral nerves.	Plexus sacralis.

No.	Common Name.	Latin or Professional Name.
53	Nerves between ribs (inter-costal.	Nervi intercostalis.
54	Network of nerves of the gullet (œsophagus.)	Plexus œsophagens.
55	Network of nerves of lungs.	Plexus pulmonalis.
56	Network of nerves of phar-ynx.	Plexus pharyngens.
57	Lower jawbone.	Os maxillare inferius.
58	Hyoid bone.	Os hyoides.
59	Shoulder bone or clavicle.	Clavicula.
60	First rib.	Costa I.
61	Second rib.	Costa II.
62	Eleventh rib.	Costa XI.
63	Transverse process of the loin backbone.	Processus transversus verte-bræ lumbalis.
64	Sacrum bone.	Os sacrum.
65	Pubis bone	Os pubis (symphysis).
66	Large cheek muscle.	Musculus zygomanticus major.
67	Lower digastric jaw muscle.	Musculus digastricus maxil-læ inferioris.
68	Chewing muscle.	Musculus masseter.
69	Salivary or parotid gland.	Glandula parotis.
70	Under jaw gland.	Glandula sub maxillaris.
71	Sterno-hyoid muscle.	Musculus sterno-hyoidens.
72	Foremost scalene muscle.	Musculus scalenus anticus.
73	Middle and posterior scal-ene muscle.	Musculus scalenus medius et posticus.
74	Midriff.	Diaphragm.
75	Square loin muscle.	Musculus quadratus lum-borum.
76	Right bronchus.	Bronchus dexter.
77	Kidney.	Renes.
78	Upper kidney gland.	Glandula supra renalis.
79	Gullet.	Œsophagus.
80	Stomach.	Stomachus.
81	Jejunum intestine.	Intestinum jejunum.
82	Colon intestine.	Intestinum colon.

No.	Common Name.	Latin or Professional Name.
83	Rectum intestine.	Intestinum rectum.
84	Bladder.	Vesica urinaria.
85	Ureter.	Ureter.
86	Procumbent gland.	Glandula prostratus.
87	Carrying vessel.	Vas deferens.
88	Spermatic cord.	Chorda spermatica.
89	Internal spermatic arteries and veins with network of internal spermatic nerves.	Arteria et vena spermatica con plexus spermaticus internus.

III. THE SENSE OF SMELL.

VERTICAL SECTION OF NASAL CAVITY.

1	Cavity in frontal bone.	Sinus frontalis ossis frontis.
2	Nasal bone.	Os nasi.
3	Sphenoidal cavity.	Sinus sphenoidalis.
4	Cribiform plate of the ethmoidal bone.	Lamina cribrosa ossis ethmoidea.
5	Upper jawbone.	Os maxillare superioris.
6	Incisive canal.	Canalis incivious.
7	Hard palate.	Palatum durum.
8	Palate molles against soft palate.	Palatum molle v. velum palatinum.
9	Tongue.	Lingua.
10	Nasal partition.	Septum nasi.
11	Posterior nasal cavity.	Posterior nares.
12	Roof of mouth.	Pharynx.
13	Tonsils.	Tonsilla.
14	Pharyngeal palate arch.	Arcus pharyngo—palatinus.
15	Olfactory nerve.	Nervus olfactorius.
16	Nasal-palate nerve of scarpa.	Nervus naso-palatinus scarpæ.
17	Incisive ganglion.	Ganglion incisivum.

IV. THE SENSE OF TASTE.

NERVES OF PALATE AND TONGUE.

No.	Common Name.	Latin or Professional Name.
1	Taste nerves.	Nervi palatini.
2	Tongue and pharynx nerve.	Nervus glosso-pharyngeus.
3	Branches of three-fold taste nerve.	Ramus gustatorius nervi trigemini.
4	Branches of No. 2.	Ramus nervi glosso-pharyngei (pro. m. glosso-palatino).
5	Upper lip.	Labium superioris.
6	Hard palate.	Palatum durum.
7	Soft palate.	Velum palatinum v. palatum molle.
8	Uvula.	Uvula.
9	Side nerve of tongue.	Arcus glosso-palatinus.
10	Arch of pharynx.	Arcus pharyngo-palatinus.
11	Tonsil.	Tonsilla.
12	Entrance to gullet.	Isthmus faucium.
13	Root of tongue.	Radix lingua.
14	Tongue.	Lingua.

V. THE SENSE OF SIGHT.

VERTICAL SECTION OF ORBIT AND GLOBE OF EYE.

1	Frontal bone.	Os frontis.
2	Upper jawbone.	Os maxillare superius.
3	Fatty matter.	Adipose tissue.
4	Frontal muscle.	Musculus frontalis.
5	Upper eyelid.	Palpebra superior.
6	Lower eyelid.	Palpebra inferior.
7	Lower oblique eye muscle.	Musculus obliquus oculi inferior.
8	Rectal eye muscle, lower.	Musculus rectus oculi inferior.

No.	Common Name.	Latin or Professional Name.
9	Rectal eye muscle, external.	Musculus rectus oculi, externus.
10	Rectal eye muscle, upper.	Musculus rectus oculi superior.
11	Upper eyelid muscle.	Musculus levator palpebræ superior.
12	Eye nerve.	Nervus opticus.
13	Conjunction of Eyelids.	Conjunctiva palpebræ.
14	Reflection of conjunction from inner surface of eyelids to globe.	
15	Conjunction of eyelids and white of eye.	Conjunctiva scleroticæ (bulbi.)
16	Conjunction of cornea.	Conjunctiva cornea.
17	Strong horny membrane forming outer part of eye.	Cornea.
18	Membrane of aqueous humor, lining anterior chamber.	
19	Anterior camera.	Camera oculi anterior
20	Posterior camera.	Camera oculi posterior.
21	Sinus of iris.	Sinus venosis iridis.
22	Sclerotic tunic.	Tunica sclerotica.
23	Crystalline lens.	Lens crystallina.
24	Ciliary body.	Corpus ciliare.
25	Vitreous body, glassy matter.	Corpus vitreum.
26	Tunic of the retina.	Tunica retina.
27	Tunic of the choroid.	Tunica choroidea.

VI. THE SENSE OF HEARING.

THE INTERNAL ORGANS OF HEARING EXPOSED WITHOUT BONY STRUCTURES.

No.	Common Name.	Latin or Professional Name.
1	External ear.	Auricula externa.
2	Anditory canal.	Meatus auditor, externus.
3	Tympanum.	Membrana tympani.
4	Hammer.	Malleus.
5	Handle of same, long.	Processus longus mallei.

No.	Common Name.	Latin or Professional Name.
6	Mannubrium of hammer.	Mannubrium mallei.
7	Anvil.	Incus.
8	Short process of same.	Processus brevis incudis.
9	Long process of same.	Processus longus incudis.
10	Orbicular ossicle.	Ossiculum orbiculare Silvii.
11	Stapes.	Stapes.
12	Vestibule.	Vestibulum.
13	Upper semicircular canal.	Canalis semicircularis superior.
14	Posterior semicircular canal.	Canalis semicircularis posterior.
15	Lower semicircular canal.	Canalis semicircularis inferior.
16	Shell, spiral cavity.	Cochlea.
17	Cupola of shell.	Cupola cochleæ.